# Advanced Network Simulations Simplified

Practical guide for wired, Wi-Fi (802.11n/ac/ax), and LTE networks using ns-3

**Dr Anil Kumar Rangisetti**

BIRMINGHAM—MUMBAI

# Advanced Network Simulations Simplified

**Group Product Manager**: Mohd Riyan Khan

**Publishing Product Manager**: Khushboo Samkaria

**Content Development Editor**: Nihar Kapadia, Divya Vijayan

**Technical Editor**: Rajat Sharma

**Copy Editor**: Safis Editing

**Project Coordinator**: Ashwin Dinesh Kharwa

**Proofreader**: Safis Editing

**Indexer**: Tejal Daruwale Soni

**Production Designer**: Joshua Misquitta

**Marketing Coordinator**: Marylou De Mello

First published: April 2023

Production reference: 1100323

Published by Packt Publishing Ltd.
Livery Place
35 Livery Street
Birmingham
B3 2PB, UK.

ISBN 978-1-80461-445-7

www.packtpub.com

*To my teachers, Dr. Bheemarjuna Reddy and Shri Badrinadh garu, for identifying my strengths, giving me wonderful opportunities to work with them, and guiding me to achieve my goals.*

*To the entire Packt team for believing in me and giving me this wonderful opportunity to publish my first book.*

*To my lovely wife, Sravani, for being a wonderful partner and supporting me in all situations.*

*– Dr Anil Kumar Rangisetti*

# Contributors

## About the author

**Dr Anil Kumar Rangisetti** received his PhD in the field of computer science and engineering from IIT Hyderabad, India. He has nearly 10 years of teaching and research experience in computer science and engineering. He currently works as an assistant professor in the Department of CSE, IIITDM Kurnool. During his career, he has worked at prestigious educational institutions and research organizations, such as IIIT Dharwad, SRM-AP, GMR, ARICENT, and IRL-Delhi. Broadly his research interests include Wi-Fi, 4G, 5G, **software-defined networking (SDN)**, **network functions virtualization** (NFV), and edge computing. He has published a number of novel research publications with IEEE, Springer, Elsevier, and Wiley in the field of a variety of networking technologies, such as **long-term evolution (LTE)**, SDN, NFV, and Wi-Fi. Besides research activities, he is interested in writing technical books on networking technologies, cloud technologies, systems, and programming languages.

# About the reviewers

**Vanlin Sathya** (IEEE member) received a PhD in computer science and engineering from the **Indian Institute of Technology** (**IIT**) Hyderabad, India, in 2016. He currently works with CTO Office at Cleona Inc., USA. Prior to this, he was a postdoctoral scholar at the University of Chicago, USA, where he worked on the issues faced in the 5G real-time coexistence test-bed when LTE-unlicensed and Wi-Fi try to coexist on the same channel. His research interests include interference management, handover in heterogeneous LTE networks, **device-to-device** (**D2D**) communication in cellular networks, cloud base station and phantom cell (LTE-B), and LTE in unlicensed and private 5G (CBRS).

**Deepanshu Khanna** is a 29-year-old information security and cybercrime consultant and a pioneer in his country. The young and dynamic personality of Deepanshu has not only assisted him in handling information security and cybercrimes but also in creating awareness about these things. He's a hacker appreciated by the Indian government, including the Ministry of Home Affairs and Defence, police departments, and many other institutes, universities, globally renowned IT firms, magazines, and newspapers. He started his career by presenting a popular hack of GRUB at HATCon. He also conducted popular research in the fields of **intruder detection software** (**IDS**) and **Advanced Intrusion Detection Environment** (**AIDE**) and demonstrated MD5 collisions and buffer overflows, among other things. His work has been published in various magazines such as pentestmag, Hakin9, e-Forensics, SD Journal, and hacker5. He has been invited as a guest speaker to public conferences such as DEF CON, ToorCon, OWASP, HATCon, H1hackz, and many other universities and institutes.

# Table of Contents

# 3

# ns-3 Key Building Blocks for Simulations          75

# Part 2: Learn, Set Up, and Evaluate Wired and Wi-Fi (802.11a/b/g/n/ac/ax) Networks

## 4

## 5

# 6

# Researching Advanced Wi-Fi Technologies – 802.11n, ac, and ax in ns-3          183

# Part 3: Learn, Set Up, and Evaluate 4G Long-Term Evolution (LTE) Networks

## 7

## 8

# 9

# Researching LTE Advanced Networks: LTE HetNets and Interference Management Using ns-3    289

# Index    327

# Other Books You May Enjoy    338

# Preface

**Network Simulator-3 (ns-3)** was created in 2006. Since then, it has been widely used by the networking community for academic and research activities. It is free, open source software and can be used under a GNU GPLv2 license. ns-3 supports a wide range of networking system simulations, such as the internet, Wi-Fi, WiMAX, ad hoc networks, and LTE. ns-3 is continuously supported by the worldwide networking community for integrating evolving networking technology simulation support and addressing all kinds of simulation issues. In this book, we used the ns-3.36 version for all simulation activities, as when we started writing this book, this was the latest version. The ns-3 official website (`www.nsnam.org`) is well maintained and offers complete documentation for all supported modules. In this book, we focus on how to use ns-3 easily and quickly for academicians, engineers, and researchers for learning basic and advanced networking technologies. There are few resources available for learning basic ns-3 simulations on the internet. In this book, we help you to learn about the following concepts easily with step-by-step procedures:

- Quickly learn about ns-3 important features: simulation logging, debugging, and tracing

- Learn how to use a wide range of ns-3 supporting nodes and network applications (TCP/UDP/HTTP), and tools for performance evaluations

- Systematic step-by-step procedures for setting up and evaluating the performance of a variety of networking topologies

The ns-3 official documentation offers example simulations and tutorials for setting up Wi-Fi, ad hoc, and LTE simulations; in this book, we will concisely introduce necessary 802.11n/ac/ax and LTE basic concepts to quickly understand the following:

- Learn how to set up and evaluate wireless ad hoc networks using a variety of placement, mobility models, and routing algorithms

- Quickly set up and learn important features of 802.11n/ac/ax networks, such as channel bonding, MIMO, and frame aggregation

- Learn how to simulate advanced 802.11ax features: resource scheduling and BSS coloring

- Learn how to set up LTE basic and advanced features: LTE networks setup, HetNets, site survey, capacity planning, radio resources management, and interference handling

There are a lot of open opportunities for network engineers and researchers in the field of Wi-Fi technology and 4G/LTE networks. This book helps you to quickly grasp important concepts and learn about advanced networking technologies confidently using interesting ns-3 hands-on activities.

# Who this book is for

Aspiring students of networking technologies, academicians, engineers, and researchers can use this book for learning about the internet, Wi-Fi, and LTE networking technologies:

- Students can use this book to thoroughly understand networking technology concepts through experiments

- Academicians can use and refer to this book for floating interesting laboratory activities, demonstrating complex networking concepts, and introducing advanced networking technologies easily

- Researchers can use this book to quickly prove their novel ideas using proof of concept results through ns-3 simulations at a low cost

- Network engineers can use this book to save their CAPEX using ns-3 simulations for debugging networking issues, site surveys, capacity planning, and learning about new networking technologies quickly

# What this book covers

*Chapter 1, Getting Started with Network Simulator-3 (ns-3)*, looks at how, before starting to use ns-3 for simulations, it is very important to learn how to use ns-3 as a simulation tool for development, testing, and evaluation activities. This chapter mainly discusses ns-3 and its supporting tools' installation for implementation and executing simulations. To make the aforementioned activities simple, ns-3 users must learn about the installation of ns-3, supporting animation tools to quickly visualize simulations, and how to write ns-3 simulation programs easily using an editor. Although the ns-3 documentation provides these details, it is difficult to follow for beginners and takes many hours to complete these activities. We explain these activities in simple steps to complete them quickly.

*Chapter 2, Monitoring, Debugging, Tracing, and Evaluating Network Topologies in ns-3*, discusses how to use ns-3 logging features to display simulation details such as nodes and their configuration details, protocol behavior, and events at various levels for informative simulations. Besides this, to understand simulation errors, exceptions, and abrupt events, ns-3 debugging and tracing features are discussed in detail. Finally, it is very important to evaluate test scenario performance in terms of throughput, delay, jitter, and packet loss metrics. In this regard, to help you, this chapter discusses how to use ns-3 FlowMonitor or evaluate test scenarios.

*Chapter 3, ns-3 Key Building Blocks for Simulations*, discusses how to set up and evaluate network topologies using ns-3 key building blocks. It discusses all the important ns-3 key building blocks, such as `Nodes`, `NodeContainers`, `NeDevices`, `InternetStack`, and a variety of channels for connecting Nodes. It also discusses how to install TCP or UDP network applications and evaluate their performance. Finally, it discusses the step-by-step procedure for setting up and evaluating simple wired or wireless LANs.

*Chapter 4, Setting Up and Evaluating CSMA/P2P LANs, Wi-Fi LANs, and the Internet*, discusses two basic LAN technologies: CSMA and P2P supported in ns-3. It mainly discusses how to set up a variety of LANs using CSMA or P2P using important network nodes, such as bridges and routers. It also discusses how to install and configure routers to connect two LANs on the internet. Finally, it introduces step-by-step procedures for how to set up Wi-Fi LANs and heterogeneous LANs.

*Chapter 5, Exploring Basic Wi-Fi Technologies, and Setting Up and Evaluating Wireless Ad Hoc Networks*, introduces the basics of Wi-Fi Nodes and protocols to set up Wi-Fi LANs and ad hoc networks using ns-3. It describes how to deploy and conduct wireless network test scenarios using placement and mobility models. In order to understand important features of Wi-Fi, it discusses Wi-Fi Nodes' operating modes, rate control algorithms, and QoS support in ns-3. Finally, it introduces wireless ad hoc networks and discusses how to set up ad hoc networks and configure a variety of routing algorithms in ns-3.

*Chapter 6, Researching Advanced Wi-Fi Technologies-802.11n, ac,and ax in ns-3*, introduces various Wi-Fi technologies (802.11a/b/g/n/ac/ax) supported in ns-3. It mainly discusses how to configure and use advanced Wi-Fi features in simulations using ns-3. It discusses how to configure frame aggregation, channel bonding, OFDM, and MIMO advanced features in Wi-Fi test scenarios using ns-3. It also discusses how to set up Wi-Fi 6 networks and evaluate resource management and spectrum reuse techniques using ns-3 simulations.

*Chapter 7, Getting Started with LTE Network Simulations Using ns-3*, introduces 4G network (LTE) simulations using ns-3. It discusses how to set up and configure LTE base stations and user equipment. Mainly, it discusses a variety of LTE network nodes, their data plane and control plane protocols, and algorithms supported in ns-3. Finally, it discusses how to set up and configure an LTE radio access network and its core network for setting up test scenarios.

*Chapter 8, Researching LTE Network Radio Resource Management and Mobility Management Using ns-3*, introduces important research aspects, such as radio resource management and mobility management in LTE networks using ns-3. It mainly focuses on how to set up an LTE network and evaluate radio resource scheduling algorithms supported in ns-3. It also discusses how to set up and conduct LTE network mobility scenarios, and evaluates LTE-supported handover algorithms in ns-3.

*Chapter 9, Researching LTE Advanced Networks: LTE HetNets and Interference Management Using ns-3*, discusses how to set up LTE advanced networks, such as heterogeneous networks, using a variety of base stations supported in ns-3. It mainly discusses how to deploy LTE HetNets by doing site surveys using the heatmaps feature of ns-3. Finally, it discusses how to configure and evaluate various LTE interference management algorithms supported in ns-3.

## To get the most out of this book

ns-3 users should have good C++ programming skills and be familiar with the Linux operating system. You should know the basics of networking technologies related to the respective simulation topics. Especially before starting Wi-Fi and LTE simulations, we recommend that you learn the basics of Wi-Fi and LTE technologies.

| Software/hardware covered in the book | Operating system requirements |
|---|---|
| ns-3.36 | Ubuntu 20.04 |
| Gnuplot, gdb, and CodeBlocks | |

# Download the color images

We also provide a PDF file that has color images of the screenshots and diagrams used in this book. You can download it here: https://packt.link/3xwd3.

# Conventions used

There are a number of text conventions used throughout this book.

`Code in text`: Indicates code words in text, database table names, folder names, filenames, file extensions, pathnames, dummy URLs, user input, and Twitter handles. Here is an example: "Open `firstanim_flowmetrics.xml` generated by your simulation program."

A block of code is set as follows:

```
NodeContainer nodes;
  nodes.Create (2);
 PointToPointHelper pointToPoint;
  pointToPoint.SetDeviceAttribute ("DataRate", StringValue
("1Gbps"));
  pointToPoint.SetChannelAttribute ("Delay", StringValue
("1ms"));
```

Any command-line input or output is written as follows:

```
$ ./ns3 run scratch/pkt_first_anim
```

**Bold**: Indicates a new term, an important word, or words that you see onscreen. For instance, words in menus or dialog boxes appear in **bold**. Here is an example: "Next, click the **Save** button."

> **Tips or important notes**
> Appear like this.

# Get in touch

Feedback from our readers is always welcome.

**General feedback**: If you have questions about any aspect of this book, email us at customercare@packtpub.com and mention the book title in the subject of your message.

**Errata**: Although we have taken every care to ensure the accuracy of our content, mistakes do happen. If you have found a mistake in this book, we would be grateful if you would report this to us. Please visit www.packtpub.com/support/errata and fill in the form.

**Piracy**: If you come across any illegal copies of our works in any form on the internet, we would be grateful if you would provide us with the location address or website name. Please contact us at copyright@packt.com with a link to the material.

**If you are interested in becoming an author**: If there is a topic that you have expertise in and you are interested in either writing or contributing to a book, please visit authors.packtpub.com

# Share your thoughts

Once you've read *Advanced Network Simulations Simplified*, we'd love to hear your thoughts! Scan the QR code below to go straight to the Amazon review page for this book and share your feedback.

https://packt.link/r/1804614459

Your review is important to us and the tech community and will help us make sure we're delivering excellent quality content.

# Download a free PDF copy of this book

Thanks for purchasing this book!

Do you like to read on the go but are unable to carry your print books everywhere?

Is your eBook purchase not compatible with the device of your choice?

Don't worry, now with every Packt book you get a DRM-free PDF version of that book at no cost.

Read anywhere, any place, on any device. Search, copy, and paste code from your favorite technical books directly into your application.

The perks don't stop there, you can get exclusive access to discounts, newsletters, and great free content in your inbox daily

Follow these simple steps to get the benefits:

1. Scan the QR code or visit the link below

https://packt.link/free-ebook/978-1-80461-445-7

2. Submit your proof of purchase

3. That's it! We'll send your free PDF and other benefits to your email directly

# Part 1:
# Exploring Network Simulator-3 (ns-3) Thoroughly and Quickly

On completion of *Part 1*, you will have learned about the installation and usage of ns-3 and its supporting tools. You will also have learned about ns-3 simulation key building blocks and important features, such as logging, debugging, tracing, and evaluating test scenarios.

This part has the following chapters:

- *Chapter 1, Getting Started with Network Simulator-3*
- *Chapter 2, Monitoring, Debugging, Tracing, and Evaluating Network Topologies in ns-3*
- *Chapter 3, ns-3 Key Building Blocks for Simulations*

# 1

# Getting Started with Network Simulator-3 (ns-3)

The main reasons for using network simulators are setting up a variety of wired or wireless networks and analyzing their protocols and application performance systematically. Conducting networking experiments is highly complex and incurs a lot of cost with real systems. **Network Simulator-3 (ns-3)** is an open source simulation platform written in C++ for conducting systematic experiments on wired and/or wireless networks. ns-3 provides important models, such as core, internet, and Ethernet-related modules, routing, applications, and flow-level statistics monitoring for setting up network topologies and conducting basic simulation experiments. Moreover, ns-3 offers important modules such as **mobility**, **placement**, **spectrum**, and **antenna** for conducting advanced network simulations related to Wi-Fi, WiMAX, and LTE networks. The ns-3 team is planning to introduce 5G simulation topics in upcoming ns-3 versions.

In general, ns-3 helps users to re-create real-time scenarios quickly and in a scalable manner. ns-3 is implemented as a discrete-event simulator to handle various events of a simulation setup. It offers a simulation engine with a scheduler that handles all types of events generated in user-written simulation experiments. ns-3's unique basic features, such as supporting logging, debugging, tracing, and computing flow-level statistics, are highly useful for conducting simulations by researchers, engineers, and students. ns-3 also supports **NetAnimator** (**NetAnim**) for visualizing ns-3 simulations. It is highly useful for beginners or advanced users to easily conduct simulations. In this chapter, we will mainly introduce the ns-3 installation procedure, and how to integrate ns-3 with `Code::Blocks` editors to quickly start ns-3 simulations. Next, we will discuss a typical ns-3 simulation program structure and how to implement and evaluate your first simulation program. Finally, we will discuss how to install and integrate NetAnim with ns-3 for visualizing ns-3 simulations. Specifically, we will discuss all the important features of NetAnim to perform systematic network simulation inspections and visualization.

In summary, in this chapter, we are going to cover the following main topics:

- Getting started with installing ns-3
- Exploring ns-3 code easily using the `Code::Blocks` editor
- Understanding a ns-3 program's structure
- Starting your first network simulation
- Playing ns-3 simulations using NetAnim

## Technical requirements

We assume you have a thorough knowledge of C++ and computer networks. Especially, for understanding Wi-Fi and LTE/4G simulations, we assume you have a basic knowledge of Wi-Fi and LTE technologies. We also recommend you revise basic network socket programming and Unix or Linux operating system commands before using the ns-3 simulator. In this book, ns-3 is set up in the working environment shown in the following table:

| Operating system | Ubuntu 20.04 |
|---|---|
| Processor | 11th Gen Intel(R) Core(TM) i5-1135G7 @ 2.40GHz 2.42 GHz |
| RAM | 8.00 GB |
| ns-3 version | 3.36 |

## Getting started with installing ns-3

There are many network simulation tools available as open source and proprietary solutions. In comparison to ns-2 and other simulation tools, ns-3 offers the following features:

- The primary changes ns-2 users observe with ns-3 is the scripting language. Also, ns-2 support is limited to simple wired and Wi-Fi network simulation only.
- ns-2 programs are written in **Object-oriented Tool Command Language** (**OTcl**), whereas ns-3 programs are written in C++ or Python.
- ns-3 is open source software and has excellent support from the ns-3 team.
- ns-3 is designed as a set of modules (internet, Wi-Fi, LTE, etc.) and users can extend existing modules and add new modules.
- ns-3 is designed to be used in the command line as well as visualization mode.
- ns-3 can be easily extended with data analysis and visualization tools.
- ns-3 can be used on Linux, macOS, or Microsoft Windows operating systems.

In this section, you will start with the first hands-on task, which is installing ns-3 successfully. Let's start by installing all necessary dependencies for ns-3:

1.  Install the core dependencies for ns-3 build essentials and compilation packages:

    ```
    sudo apt-get update
    sudo apt install build-essential libsqlite3-dev libboost-
    all-dev libssl-dev git python3-setuptools castxml
    ```

2.  Install the Python dependencies for ns-3 Python bindings:

    ```
    sudo apt install girl.2-goocanvas-2.0 girl.2-gtk-3.0
    libgirepository1.0-dev python3-dev python3-gi python3-gi-
    cairo python3-pip python3-pygraphviz python3-pygccxml
    ```

3.  Install the dependencies for ns-3 features support, such as NetAnimator, gdb, and valgrind:

    ```
    sudo apt install g++ pkg-config sqlite3 qt5-default
    mercurial ipython3 openmpi-bin openmpi-common openmpi-
    doc libopenmpi-dev autoconf cvs bzr unrar gdb valgrind
    uncrustify doxygen graphviz imagemagick python3-sphinx
    dia tcpdump libxml2 libxml2-dev cmake libc6-dev libc6-
    dev-i386 libclang-6.0-dev llvm-6.0-dev automake
    ```

    Next, download the latest ns-3 version and follow the steps for installing it:

    ```
    $ wget -c https://www.nsnam.org/releases/
    ns-allinone-3.36.tar.bz2
    ```

4.  Unzip or extract the downloaded file:

    ```
    $ tar -xvjf ns-allinone-3.36.tar.bz2
    ```

5.  Move to the following directory:

    ```
    $ cd ns-allinone-3.36/ns-3.36
    ```

6.  Next, configure the ns-3 modules using the build system. The ns3 command (available in the ns-allinone-3.36/ns-3.36 folder) makes use of a Python wrapper around CMake. It is similar to Waf in earlier ns-3 versions. The following command configures ns-3 modules with all example and test simulation programs:

    ```
    $ ./ns3 configure --enable-examples --enable-tests
    ```

After running this command, you can verify the list of modules configured in *Figure 1.1*:

```
Modules configured to be built:
antenna                aodv                   applications
bridge                 buildings              config-store
core                   csma                   csma-layout
dsdv                   dsr                    energy
fd-net-device          flow-monitor           internet
internet-apps          lr-wpan                lte
mesh                   mobility               netanim
network                nix-vector-routing     olsr
point-to-point         point-to-point-layout  propagation
sixlowpan              spectrum               stats
tap-bridge             test                   topology-read
traffic-control        uan                    virtual-net-device
wave                   wifi                   wimax

Modules that cannot be built:
brite                  click                  mpi
openflow               visualizer
```

Figure 1.1 – The list of ns-3 modules configured

7. Next, build all the configured ns-3 modules using the following command. This command takes a while to complete:

   ```
   $ ./ns3 build
   ```

   After executing the command, we can observe the list of modules linking successfully, as shown in *Figure 1.2*:

   ```
   Scanning dependencies of target test-runner
   [ 99%] Building CXX object utils/CMakeFiles/test-runner.dir/test-runner.cc.o
   [100%] Linking CXX executable ../../build/utils/ns3.36-test-runner-default
   Finished executing the following commands:
   cd cmake-cache; cmake --build . -j 11 ; cd ..
   ```

   Figure 1.2 – ns-3 build process

8. Finally, you can check your installation by running all unit test cases provided in ns-3 modules:

   ```
   $ ./test.py
   ```

   After executing this command, all test cases should either pass or be skipped. Then, a successfully installed NS-3.36 simulator is available (refer to *Figure 1.3*):

   ```
   [729/731] PASS: Example src/wifi/examples/wifi-manager-example --wifiManager=MinstrelHt
   ShortGuardInterval=3200 --serverNss=4 --clientNss=4 --stepTime=0.1
   [730/731] PASS: Example src/wifi/examples/wifi-manager-example --wifiManager=MinstrelHt
   ShortGuardInterval=1600 --serverNss=4 --clientNss=4 --stepTime=0.1
   [731/731] PASS: Example src/wifi/examples/wifi-manager-example --wifiManager=MinstrelHt
   ShortGuardInterval=3200 --serverNss=4 --clientNss=4 --stepTime=0.1
   731 of 731 tests passed (731 passed, 0 skipped, 0 failed, 0 crashed, 0 valgrind errors)
   ```

   Figure 1.3 – ns-3 test cases execution

9. After the successful build of ns-3, you have many example simulation programs ready to execute. Try running a first example simulation with the following command and observe the output:

   ```
   $ ./ns3 run first
   ```

Now, we are executing an example ns-3 simulation, `first.cc`. Refer to the execution results in *Figure 1.4*:

```
At time +2s client sent 1024 bytes to 10.1.1.2 port 9
At time +2.00369s server received 1024 bytes from 10.1.1.1 port 49153
At time +2.00369s server sent 1024 bytes to 10.1.1.1 port 49153
At time +2.00737s client received 1024 bytes from 10.1.1.2 port 9
```

Figure 1.4 – The first.cc simulation execution results

Congratulations! We have installed ns-3 successfully and tested a sample simulation. In the next section, we are going to learn how to write an ns-3 simulation program easily using `Code::Blocks` editor features. Thanks to the entire ns-3 team for providing detailed documentation on their website (`https://www.nsnam.org`) related to installing ns-3 with all features, tutorials, examples, and designs of various modules. We recommend you to go through the website for more details.

# Exploring ns-3 code easily using the Code::Blocks editor

After installing ns-3, the most exciting task is writing the first simulation program. Thanks to ns-3 for providing various tutorials and example programs in various modules. We strongly recommend ns-3 beginners start by copying the example programs and understanding their source code using editors such as `Code::Blocks` and Visual Studio. This will help users to understand key packages, classes, fields, and member functions used in various programs. Moreover, while writing or editing programs, the IntelliSense features of the editors will help users to learn all the basic and unique features supported by ns-3 classes such as fields, constructors, member functions, attributes, callbacks, and trace sources. Hence, users can write their simulation programs easily and quickly by using the IntelliSense and auto-completion features of the editors. It saves a lot of time for users as they can use classes and all its features without searching the internet, ns-3 documentation, and example programs thoroughly. ns-3 offers support for the following editors, based on operating system: `Code::Blocks` for Ubuntu and Visual Studio for Windows. In this book, we use the `Code::Blocks` editor.

## Installing and configuring Code::Blocks for ns-3

Let's start by installing `Code::Blocks` and creating an ns-3 `Code::Blocks` project:

```
$ apt-get install codeblocks
```

By default, the `Code::Blocks` editor does not support ns-3 projects. Hence, it is necessary to create a ns-3 `Code::Blocks` project using the following command from the ns-3 installed directory (`ns-allinone3.36/ns-3.36`):

```
$./ns3 configure -G"CodeBlocks - Unix Makefiles" --enable-
examples
```

The preceding command creates an NS3.cbp Code::Blocks project file inside cmake-cache. This is a Code::Blocks project file that can be opened by the IDE. Now, open the Code::Blocks editor. Then, move to **File** | **Open** | ns-allinone3.36 | ns-3.36 | cmake-cache and select the NS3.cbp file. It opens the NS3 project, as shown in *Figure 1.5*:

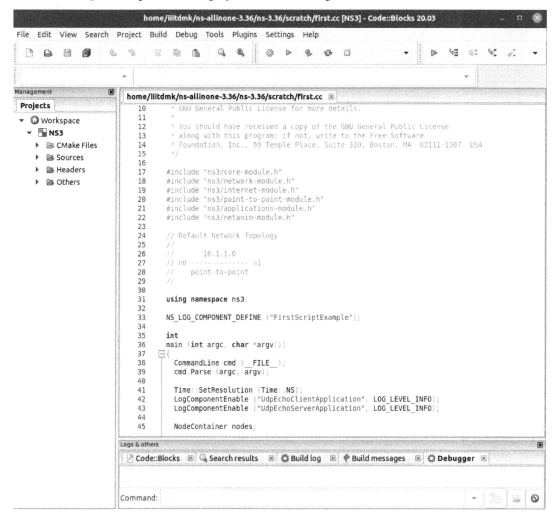

Figure 1.5 – The ns-3 project opens in the Code::Blocks editor

Next, quickly check the **project properties** to run a sample first simulation. It opens the window shown in *Figure 1.6*. In the window, no project settings need to be changed but make sure you check the **Build targets** tab. Under the selected **Build targets** options panel, select the **Type** drop-down menu, then select **Console application** and click OK.

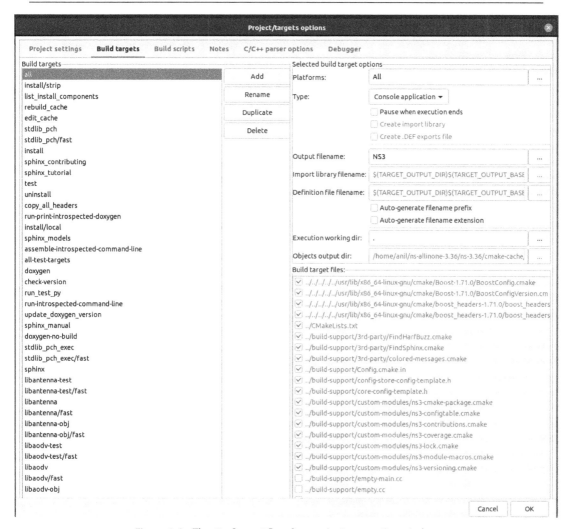

Figure 1.6 – The Code::Blocks project properties window

Next, execute a first.cc simulation. Go to the **Build** menu, then **Select target**. It opens the following window (refer to *Figure 1.7*). Type scrach_first to run your first simulation. Select the **Run** button from the editor to execute the first simulation:

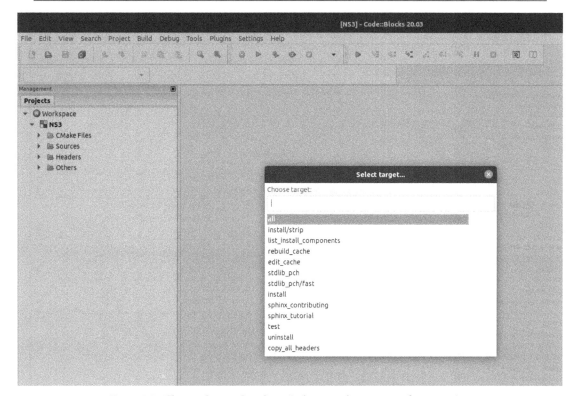

Figure 1.7 – The Code::Blocks window to select a target for execution

By clicking on the **Run** button, it successfully executes the first simulation, as shown in *Figure 1.8*. It displays output in a new window. This was just a demonstration of how to select a simulation program from the Code::Blocks editor and execute it:

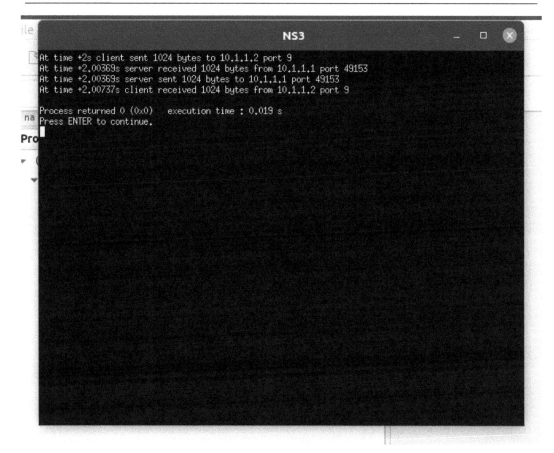

Figure 1.8 – The first.cc simulation execution results

Next, we are going to explore the ns-3 module's source code using the `Code::Blocks` editor.

## Exploring various ns-3 modules' source code

Under the `Code::Blocks` editor **Workspace**, you can see NS3 project folders (refer to *Figure 1.9*). Select **Sources** and traverse to the `src` folder. Then, you can see the following ns-3 modules list, including **antenna**, **aodv**, and **applications**:

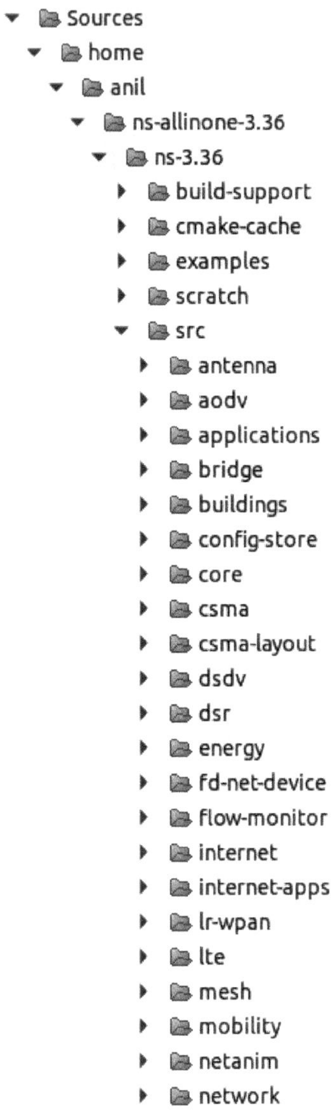

```
▼  📁 Sources
   ▼  📁 home
      ▼  📁 anil
         ▼  📁 ns-allinone-3.36
            ▼  📁 ns-3.36
               ▶  📁 build-support
               ▶  📁 cmake-cache
               ▶  📁 examples
               ▶  📁 scratch
               ▼  📁 src
                  ▶  📁 antenna
                  ▶  📁 aodv
                  ▶  📁 applications
                  ▶  📁 bridge
                  ▶  📁 buildings
                  ▶  📁 config-store
                  ▶  📁 core
                  ▶  📁 csma
                  ▶  📁 csma-layout
                  ▶  📁 dsdv
                  ▶  📁 dsr
                  ▶  📁 energy
                  ▶  📁 fd-net-device
                  ▶  📁 flow-monitor
                  ▶  📁 internet
                  ▶  📁 internet-apps
                  ▶  📁 lr-wpan
                  ▶  📁 lte
                  ▶  📁 mesh
                  ▶  📁 mobility
                  ▶  📁 netanim
                  ▶  📁 network
```

Figure 1.9 – The Code::Blocks editor Workspace

In ns-3, every module implementation follows a hierarchy. For an example, we will explore the **lte** module and its underlying folders (refer to *Figure 1.10*):

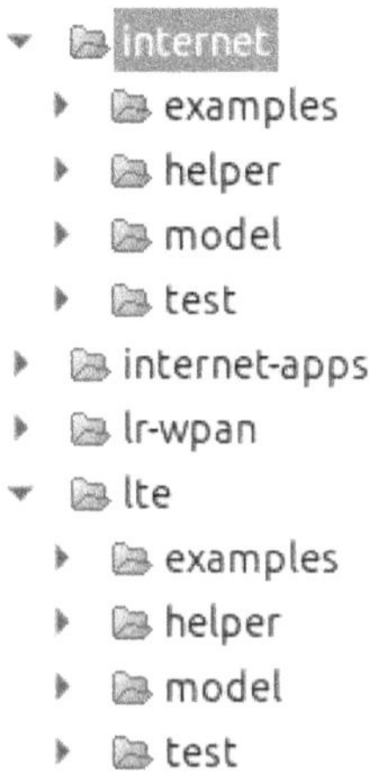

Figure 1.10 – An ns-3 example module (lte) and its common folders

As we can observe, there are four folder names – **examples**, **helper**, **model**, and **test**:

- The **model** folder contains all LTE protocols', architecture's, interfaces', and algorithms' source code. Developers can explore the **model** folder and its source files to make changes in LTE core-level source code. For example, developers use `lte-enb-mac.cc` to change any LTE **base station** (**BS**)'s **medium access control** (**MAC**) protocol changes.

- The **examples** folder contains sample LTE simulation programs, which can be used by end users to understand how to write LTE simulations. For example, users can explore `lena-x2-handover.cc` to understand how to write a mobility scenario in LTE networks.

- The **helper** folder contains source code useful for setting up simulation programs. To set up LTE network simulation, end users use helper classes of `lte-helper.cc` and `epc-helper.cc`. Usually, the **helper** folder contains all helper classes for setting up network architectures, interfaces, and protocols.

- The **test** folder contains the source code of various unit test cases for the module. For example, the **lte** test folder contains all of the various test programs for checking LTE nodes' behavior, protocols, interfaces, and algorithms.

All ns-3 modules follow the same folder hierarchy. Hence, it is important for developers and users to understand its hierarchy to make any changes.

Next, we'll see how the `Code::Blocks` editor is useful for writing ns-3 simulation programs. The `Code::Blocks` IntelliSense and auto-completion features are also available for ns-3 programs as they are for C++ programs. Let's open `first.cc` and see how useful the `Code::Blocks` editor is for making any changes or writing new lines of code. As we are new to ns-3 programming, it is important to understand and remember various classes' syntax and semantics. Using a regular editor will not help you in terms of auto code completion and IntelliSense features. As ns-3 is configured as a `Code::Blocks` project, it is possible to use `Code::Blocks` editor features to write ns-3 programs like C or C++ programs.

You can use the auto code completion feature by typing the first few characters of an ns-3 class, field, or member function. For example, if you type `NodeC`, the editor will suggest the **NodeContainer** class. More importantly, due to the IntelliSense feature, we can view a complete description of any class (refer to *Figure 1.11*), field, or member function before including it in our code:

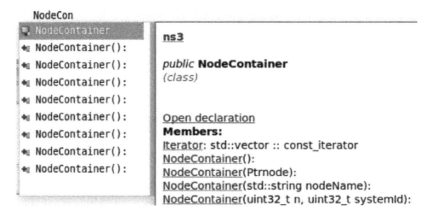

Figure 1.11 – Inspect the ns-3 NodeContainer class

For example, while typing `NodeC`, it opens a class description in a window. We can click **Open declaration** to see the entire class definition or we can scroll down the window to see its members, fields, iterators, constructors, attributes, and so on. As a beginner, it is difficult to write an ns-3 program from scratch, hence we recommend that you use example programs, understand them, and modify them using `Code::Blocks` or any ns-3-supported editor.

That's great. We have learned how to integrate ns-3 as a `Code::Blocks` project and explored ns-3 modules' source code structure. In the next section, we will start understanding an ns-3 simulation program's structure to write simulation programs.

## Understanding an ns-3 program's structure

Before writing any simulation program, it is important to understand the ns-3 simulation program structure. By doing so, it is possible to avoid common mistakes and save a lot of time in debugging your simulation issues. Here, we describe how to create the basic and advanced ns-3 simulation program structures, as follows:

1.  Import all necessary ns-3 packages.

2.  Define the logging component for your program. Your simulation program allows you to include necessary log statements, such as information, warnings, errors, exceptions, assertions, and debug details. While running a simulation, you can enable or disable any of the log statements.

3. Set up a simulation topology and configure it. Before writing your simulation program, it is recommended to draw your simulation network topology:

   A. Create **nodes** using the ns-3 `NodeContainer` class. In ns-3, hosts and other network equipment are referred to as **nodes**, and users can install the necessary protocols to configure them as a specific piece of networking equipment. Identify the type of nodes, such as hosts, routers, switches, **access points** (**APs**), and BSs, involved in your topology and create the necessary number of nodes for each type using the ns-3 `NodeContainer` class.

   B. Connect nodes using suitable communication channels and install the necessary communication protocols to get the `NetDevices` (**network interface cards** (**NIC**)) associated with the nodes. For example, you'll need to collect the necessary wired or wireless/mobile network devices using the ns-3 `NetDeviceContainer` class.

   C. Install the **Transmission Control Protocol** (**TCP**)/**Internet Protocol** (**IP**) on nodes using the ns-3 `InternetStackHelper` class. Hence, we can assign IP addresses and install routing protocols.

   D. Configure IP addresses. We need to identify and create types of network addresses as required by setting suitable network IDs and subnet masks. Then, we can configure the respective IP address to respective `NetDevices` using the ns-3 `NetDeviceContainer` class.

   E. Configure routing protocols to interconnect different networks involved in your topology.

4. Next, identify the necessary network applications to install and run on various nodes:

   A. Identify on which nodes you need to install your client or server applications, then configure suitable IP addresses and port numbers for those client and server applications.

   B. Configure client and server application traffic flow characteristics, such as packet size, packet interval, and number of packets.

   C. Install the network applications on the selected nodes.

   D. Configure the start and stop timings of your applications.

5. Install `FlowMonitor` (an ns-3 class) on all nodes to collect all flow statistics. Although it is not necessary to install a `FlowMonitor` application on nodes, it is important to analyze your network application's performance on the simulation topology in terms of flow level throughput, delay, packet loss, and jitter metrics.

6. Configure packet capturing and traces on nodes (this is optional). This helps you to collect a variety of packets exchanged across your simulation topology and possibly to inspect packet details using the `wireshark` or `tcpdump` packet analysis tools.

7.  Configure the Animation Interface (this is optional). It helps you to view the following simulation execution and results using the NetAnim visualization tool:

    A.  Enable various packets exchange among nodes tracking feature.

    B.  Enable the routing tables tracking feature.

    C.  Enable the counter tables tracking feature.

    D.  Enable flow monitor statistics capturing.

8.  Configure the stopping time of your simulation using `Simulator::Stop()`. It is necessary to stop the simulation when you use a flow monitor; otherwise, your simulation never ends.

9.  Start running your simulation using `Simulator::Run()`. This statement starts your simulation execution.

10. Convert the flow monitor results into an XML file to visualize using NetAnim.

11. Clean up your simulation resources using `Simulator::Destroy()`. It is important to clean up your simulation resources such as memory, devices, and files.

We now understand a typical ns-3 simulation program structure. This common structure is suitable for writing any wired, wireless, or mobile networking topology simulation. With this knowledge, in the next section, we will start writing our first simulation program (`pkt_first.cc`) by referring to the existing tutorial/`first.cc` simulation program.

## Starting your first network simulation

In ns-3 topologies, host or any network equipment are referred to as **nodes** with unique identifiers. We can view a node as a computing device. Based on our requirements, we can install necessary applications, and wired, wireless, or IP stacks to enable a given node as suitable network equipment or a suitable host or server. Besides this, to connect various nodes, ns-3 offers a variety of wired and wireless channels. In ns-3, nodes and channels are implemented using C++ classes. Usually, ns-3 simulation programs are written in C++ and saved with the `.cc` extension under the `ns-allinone-3.36/ns-3.36/scratch/` folder. You can write an ns-3 program in any text editor or `Code::Blocks` like any other C++ program. Now, we will discuss our first simulation program (`pkt_first.cc`) for simulating a simple **point-to-point** (**P2P**) network topology.

## First simulation implementation activity using ns-3

What do you want to learn and achieve in your first network simulation activity? Let's look at some suggestions:

- This is how to set up the following simple P2P network (shown in comments), which connects two hosts. In ns-3, the P2P module is a P2P data link available for connecting exactly two devices only using a `PointToPointHelper` channel class:

```
// Simple Point to Point Network
//          192.168.1.0
// Node 0 -------------- Node 1
//    point-to-point channel
// (Speed: 1Gbps, Propagation Delay: 1ms)
// Install User Datagram Protocol (UDP) Echo Client on
Node-0
// Install UDP Echo Server on Node-1
```

- Install the UDP echo client and server applications on Node-0 and Node-1 respectively and configure the following UDP echo client application traffic characteristics:

  - `MaxPackets`: Maximum packets to be sent (e.g., 10).

  - `Interval`: At what speed the echo client application should send a packet. For example, setting `Interval` to `0.1` means 10 packets are sent per second.

  - `PacketSize`: Configure each packet size in terms of bytes (e.g., 1,024).

- Monitor how the UDP echo client and server exchange packets in the network.

- View the actual packets processed at Node-0 and Node-1.

- Generate an animation XML file that can be used with NetAnim later.

- Analyze UDP flow performance in terms of throughput, delay, jitter, and packet loss.

Now, we are going to explain step-by-step instructions for how to solve the preceding simulation activity using an ns-3 program, which will be saved under `pkt_first.cc` in the `ns-allinone-3.36/ns-3.36/scratch/` folder:

1. First, import all necessary ns-3 packages:

    A. `core-module` for accessing all core features of ns-3, such as simulator activities in terms of handling events, starting and stopping simulation execution, logging, tracing, and attributes.

B.  `network-module` for accessing network socket features, packet representation, and so on.

C.  `internet-module` for accessing and installing TCP/IP protocols and `applications-module` for installing various TCP/UDP applications on nodes.

D.  Next, specific to our simulation, we use `point-to-point-module` to access and use the P2P channel.

E.  Finally, we use `netanim-module` to enable animation support and use `flow-monitor-module` to collect flow-level statistics, such as throughput, delay, jitter, and packet loss:

```
#include "ns3/core-module.h"
#include "ns3/network-module.h"
#include "ns3/internet-module.h"
#include "ns3/applications-module.h"
#include "ns3/point-to-point-module.h"
#include "ns3/netanim-module.h"
#include "ns3/flow-monitor-module.h"
using namespace ns-3
```

2.  In ns-3, the actual simulation execution starts with the `main()` function. Hence, we need to write our actual code in the `main()` function. We write the following relevant code in `main()`:

A.  As per the topology diagram given in our simulation activity, we start by first creating two hosts as two nodes using `NodeContainer`.

B.  Then, we create and configure the P2P channel as per our requirements (1 Gbps speed, 1 ms delay) using `PointToPointHelper`:

```
NodeContainer nodes;
nodes.Create (2);
PointToPointHelper pointToPoint;
pointToPoint.SetDeviceAttribute ("DataRate",
StringValue ("1Gbps"));
pointToPoint.SetChannelAttribute ("Delay", StringValue
("1ms"));
```

C.  Next, install the P2P protocol on the nodes, which connects them using the P2P channel. As a result, it returns two devices to the `NetDeviceContainer` devices: the first device is the Node-0 NIC and the second device is the Node-1 NIC:

```
NetDeviceContainer devices;
devices = pointToPoint.Install (nodes);
```

D.  Now, we install TCP/IP protocols on nodes. Then, we configure the IP addresses for the `NetDeviceContainer` devices under network ID `192.168.1.0`. As a result, Node-0 gets `192.168.1.1` and Node-1 gets `192.168.1.2`:

```
InternetStackHelper stack;
stack.Install (nodes);
Ipv4AddressHelper address;
address.SetBase ("192.168.1.0", "255.255.255.0");
Ipv4InterfaceContainer interfaces = address.Assign
(devices);
```

3.  Create traffic applications and install them on specific ns-3 nodes:

A.  First, create a UDP echo server application and install it on Node-1. Configure its start and stop timings:

```
uint64_t port = 9;
UdpEchoServerHelper echoServer (port);
ApplicationContainer serverApps = echoServer.Install
(nodes.Get (1));
serverApps.Start (Seconds (1.0));
serverApps.Stop (Seconds (10.0));
```

B.  Next, create a UDP echo client application and install it on Node-0. Configure its traffic characteristics, such as the maximum number of packets to be sent, packet-sending interval, and packet size. Then, configure UDP echo client start and stop timings as per the number of packets to be sent:

```
UdpEchoClientHelper echoClient (interfaces.GetAddress
(1), port);
echoClient.SetAttribute ("MaxPackets", UintegerValue
(10000000));
echoClient.SetAttribute ("Interval", TimeValue (Seconds
(0.001)));
echoClient.SetAttribute ("PacketSize", UintegerValue
(1024));
ApplicationContainer clientApps = echoClient.Install
(nodes.Get (0));
clientApps.Start (Seconds (2.0));
clientApps.Stop (Seconds (10.0));
```

4.  Install `FlowMonitor` on all nodes to collect flow-level statistics. By doing so, at the end of the simulation, we can view and analyze each traffic flow's throughput, delay, jitter, and packet loss:

```
Ptr<FlowMonitor> flowmon;
FlowMonitorHelper flowmonHelper;
flowmon = flowmonHelper.InstallAll ();
```

5.  Enable packet capturing on all nodes. In doing so, at the end of the simulation, it is possible to view the `pcap` files generated using the `wireshark` or `tcpdump` tools:

```
pointToPoint.EnablePcapAll ("pkt_first");
```

6.  Configure the animation interface. By doing so, after completing the simulation execution, an animation XML file will be generated. The XML file can be viewed using NetAnim and visualizing the simulation execution:

```
AnimationInterface anim ("first_animation.xml");
anim.EnablePacketMetadata (); // Optional
```

7.  Configure the stopping time of your simulation using `Simulator::Stop()` and set its stop time value to be greater than the client and server applications' stop timer values:

```
Simulator::Stop (Seconds(11.0));
```

8.  Start running your simulation using `Simulator::Run()`:

```
Simulator::Run ();
```

9.  Convert the flow monitor results into an XML file to parse and view it at the end of the simulation:

```
flowmon->SerializeToXmlFile ("first_flowmetrics.xml",
true, true);
```

10. Clean up your simulation resources using `Simulator::Destroy()`:

```
Simulator::Destroy ();
```

Let's see the complete first simulation program (`pkt_first.cc`):

```
#include "ns3/core-module.h"
#include "ns3/network-module.h"
#include "ns3/internet-module.h"
#include "ns3/point-to-point-module.h"
#include "ns3/applications-module.h"
#include "ns3/netanim-module.h"
```

```
#include "ns3/flow-monitor-module.h"
using namespace ns3;
NS_LOG_COMPONENT_DEFINE ("FirstScriptExample");
int
main (int argc, char *argv[])
{
 LogComponentEnable ("UdpEchoClientApplication", LOG_LEVEL_
INFO);
  LogComponentEnable ("UdpEchoServerApplication", LOG_LEVEL_
INFO);
  NodeContainer nodes;
  nodes.Create (2);
  PointToPointHelper pointToPoint;
  pointToPoint.SetDeviceAttribute ("DataRate", StringValue
("1Gbps"));
  pointToPoint.SetChannelAttribute ("Delay", StringValue
("1ms"));

  NetDeviceContainer devices;
  devices = pointToPoint.Install (nodes);
  InternetStackHelper stack;
  stack.Install (nodes);
  Ipv4AddressHelper address;
  address.SetBase ("192.168.1.0", "255.255.255.0");
  Ipv4InterfaceContainer interfaces = address.Assign (devices);

  uint64_t port = 9;
  UdpEchoServerHelper echoServer (port);
  ApplicationContainer serverApps = echoServer.Install (nodes.
Get (1));
  serverApps.Start (Seconds (1.0));
  serverApps.Stop (Seconds (10.0));

  UdpEchoClientHelper echoClient (interfaces.GetAddress (1),
port);
  echoClient.SetAttribute ("MaxPackets", UintegerValue (10));
```

```
  echoClient.SetAttribute ("Interval", TimeValue (Seconds
(0.1)));
  echoClient.SetAttribute ("PacketSize", UintegerValue (1024));
  ApplicationContainer clientApps = echoClient.Install (nodes.
Get (0));
  clientApps.Start (Seconds (2.0));
  clientApps.Stop (Seconds (10.0));

  Ptr<FlowMonitor> flowmon;
  FlowMonitorHelper flowmonHelper;
  flowmon = flowmonHelper.InstallAll ();
 pointToPoint.EnablePcapAll ("pkt_first");
 AnimationInterface anim ("first_animation.xml");
  anim.EnablePacketMetadata (); // Optional

  Simulator::Stop (Seconds(11.0));
  Simulator::Run ();
  flowmon->SerializeToXmlFile ("first_flowmetrics.xml", true,
true);
  Simulator::Destroy ();
}
```

That's great. We have completed our first simulation implementation. Let's evaluate and analyze our first simulation thoroughly in the next section.

## First simulation evaluation using ns-3

Let's run our first simulation program using the following command. As pkt_first.cc is saved in the scratch folder, first move it to the ns-allinone-3.36/ns-3.36 folder and run the following command:

```
$./ns3 run scratch/pkt_first
```

After executing the command, let's observe the simulation results in *Figure 1.12*:

```
At time +2s client sent 1024 bytes to 192.168.1.2 port 9
At time +2.00101s server received 1024 bytes from 192.168.1.1 port 49153
At time +2.00101s server sent 1024 bytes to 192.168.1.1 port 49153
At time +2.00202s client received 1024 bytes from 192.168.1.2 port 9
At time +2.1s client sent 1024 bytes to 192.168.1.2 port 9
At time +2.10101s server received 1024 bytes from 192.168.1.1 port 49153
At time +2.10101s server sent 1024 bytes to 192.168.1.1 port 49153
At time +2.10202s client received 1024 bytes from 192.168.1.2 port 9
At time +2.2s client sent 1024 bytes to 192.168.1.2 port 9
At time +2.20101s server received 1024 bytes from 192.168.1.1 port 49153
At time +2.20101s server sent 1024 bytes to 192.168.1.1 port 49153
At time +2.20202s client received 1024 bytes from 192.168.1.2 port 9
At time +2.3s client sent 1024 bytes to 192.168.1.2 port 9
At time +2.30101s server received 1024 bytes from 192.168.1.1 port 49153
At time +2.30101s server sent 1024 bytes to 192.168.1.1 port 49153
At time +2.30202s client received 1024 bytes from 192.168.1.2 port 9
At time +2.4s client sent 1024 bytes to 192.168.1.2 port 9
At time +2.40101s server received 1024 bytes from 192.168.1.1 port 49153
At time +2.40101s server sent 1024 bytes to 192.168.1.1 port 49153
At time +2.40202s client received 1024 bytes from 192.168.1.2 port 9
At time +2.5s client sent 1024 bytes to 192.168.1.2 port 9
At time +2.50101s server received 1024 bytes from 192.168.1.1 port 49153
At time +2.50101s server sent 1024 bytes to 192.168.1.1 port 49153
At time +2.50202s client received 1024 bytes from 192.168.1.2 port 9
At time +2.6s client sent 1024 bytes to 192.168.1.2 port 9
At time +2.60101s server received 1024 bytes from 192.168.1.1 port 49153
At time +2.60101s server sent 1024 bytes to 192.168.1.1 port 49153
At time +2.60202s client received 1024 bytes from 192.168.1.2 port 9
At time +2.7s client sent 1024 bytes to 192.168.1.2 port 9
At time +2.70101s server received 1024 bytes from 192.168.1.1 port 49153
At time +2.70101s server sent 1024 bytes to 192.168.1.1 port 49153
At time +2.70202s client received 1024 bytes from 192.168.1.2 port 9
At time +2.8s client sent 1024 bytes to 192.168.1.2 port 9
At time +2.80101s server received 1024 bytes from 192.168.1.1 port 49153
At time +2.80101s server sent 1024 bytes to 192.168.1.1 port 49153
At time +2.80202s client received 1024 bytes from 192.168.1.2 port 9
At time +2.9s client sent 1024 bytes to 192.168.1.2 port 9
At time +2.90101s server received 1024 bytes from 192.168.1.1 port 49153
At time +2.90101s server sent 1024 bytes to 192.168.1.1 port 49153
At time +2.90202s client received 1024 bytes from 192.168.1.2 port 9
```

Figure 1.12 – The pkt_first.cc simulation execution results

After running the pkt_first.cc simulation, the following details need to be observed.

Monitor the simulation execution, including things such as how the two nodes exchange UDP echo requests and reply packets. Since we enabled the log components of the UDP echo server and client applications, it is possible to view the various events generated by the UDP applications used in our simulation. We can observe a few important details, such as at what time the first packet is generated by the client (at 2 seconds), at what time the server receives a packet and when it sends a reply for the packet, and the total number of packets exchanged between the client and server.

Check the list of files generated in *Figure 1.13*:

```
171224 Jul 13 14:43 pkt_first-1-0.pcap
171224 Jul 13 14:43 pkt_first-0-0.pcap
  2784 Jul 13 14:43 first_flowmetrics.xml
 59939 Jul 13 14:43 first_animation.xml
```

Figure 1.13 – The list of files generated at end of the simulation

We can observe a total of four files are generated at the end of the simulation.

The first are PCAP files. As we are interested in viewing packets processed at Node 0 and 1, we enabled packet capturing on those nodes using `pointToPoint.EnablePcapAll ("pkt_first");` in our simulation program. It generates two files whose names start with `pkt_first` followed by Node and its NIC numbers; for example, `pkt_first_0_0.pcap` corresponds to Node-0 and NIC-0, and `pkt_first_1_0.pcap` corresponds to Node-1 and NIC-0. We can open these `.pcap` files using wireshark (refer to *Figure 1.14*) or the `tcpdump` tool to view the file contents:

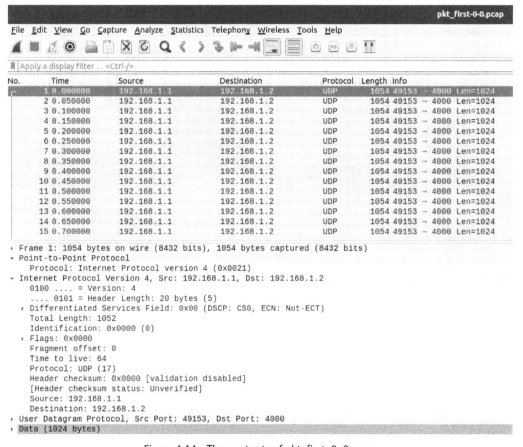

Figure 1.14 – The contents of pkt_first_0_0.pcap

Then we have two XML files. One file is `first_animation.xml`. This file can be used to visualize the simulation execution using the NetAnim tool. Another XML file is `first_flowmetrics.xml`. This file is useful for viewing flow statistics such as throughput, delay, jitter, and packet loss.

Let's validate our simulation results with the ns-3 `FlowMonitor` using the following parsing commands.

First copy `ns-allinone-3.36/ns-3.36/src/flow-monitor/examples/flowmon-parse-results.py` to the `scratch` folder. Execute the following command and observe the following results in *Figure 1.15*:

```
python3 scratch/flowmon-parse-results.py first_flowmetrics.xml
```

In our simulation, the UDP client application sends 10 packets of 1,024 bytes at a rate of one per second (10*1,024*8) to the echo server application. Hence, for both flows, we see the same TX and RX bitrates (93 kbps) in the flow monitor results shown in *Figure 1.15*. These results match our configured flow rate per second. Similarly, the average delay (1.01 ms) and packet loss (0%) can be verified:

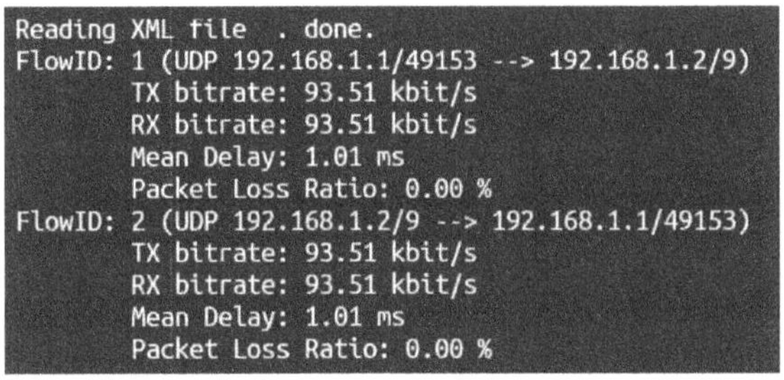

```
Reading XML file  . done.
FlowID: 1 (UDP 192.168.1.1/49153 --> 192.168.1.2/9)
        TX bitrate: 93.51 kbit/s
        RX bitrate: 93.51 kbit/s
        Mean Delay: 1.01 ms
        Packet Loss Ratio: 0.00 %
FlowID: 2 (UDP 192.168.1.2/9 --> 192.168.1.1/49153)
        TX bitrate: 93.51 kbit/s
        RX bitrate: 93.51 kbit/s
        Mean Delay: 1.01 ms
        Packet Loss Ratio: 0.00 %
```

Figure 1.15 – Flow monitor results

You can try changing `MaxPackets`, `Interval`, and `PacketSize` to observe different results.

The following are points to remember and check while implementing simulation programs:

- Check whether you imported all necessary packages.

- A few points to bear in mind while creating a topology are the following:

  - The node index starts at 0.

  - After connecting all nodes using channels, a node can have multiple network devices (or NICs). The NIC index also starts at 0.

  - IP addresses should be assigned to nodes only after installing the necessary protocols and internet stack. From a subnet, the first IP will be assigned to the first node, the second IP assigned to the second node, and so on.

  - Routing protocols should be configured on nodes only after IP address assignment.

- It is necessary to confirm on which nodes the client or server applications are installed and check whether the client is configured with the correct server's socket address (server IP and port number). This helps avoid common mistakes in installing applications.

- Before running a simulation, it is also necessary to check application traffic characteristics such as the packet size, number of packets, packet intervals, and start and stop timings of respective client and server applications. This helps avoid common mistakes when computing flow-level metrics such as throughput, delay, jitter, and packet loss.

- It is good practice to install `FlowMonitor` in your simulation setup to analyze your simulation results. When using `FlowMonitor`, the following should be done:

  - It is necessary to use the `Simulator::Stop()` statement before `Simulator::Run()` to break infinite events generated by `FlowMonitor`.

  - After the `Simulator:: Run()` statement, it is necessary to use `SerializeToXmlFile()` to view results in the NetAnim visualizer.

- By default, logging, packet capturing, and tracing features are not enabled.

- In cases where users are interested in visualizing simulation results, it is necessary to write all suitable NetAnim statements in the program.

- Set the simulator stop time to a greater value than the client and server applications' stop timer values.

In this section, we have learned how to write a simulation program using various ns-3 classes, and we also learned about important things to observe during and after a simulation execution. In the next section, we start the most fun and long-awaited ns-3 session using NetAnim.

## Playing ns-3 simulations using NetAnim

One of the interesting and very useful features of ns-3 is the support of NetAnim. NetAnim makes executing simulations more interesting. It allows you to visualize the entire simulation execution in a bird's eye view in terms of inspecting a variety of nodes, their configuration details, placement models, mobility models, and packet flows. Besides visualizing simulation execution using NetAnim, we can do the following interesting activities:

- Monitor and inspect various protocol packets exchanged among network nodes during a traffic flow

- Inspect network node configuration details such as IP, MAC, routing tables, and counters

- View flow-level statistics such as throughput, delay, jitter, and packet loss

In the next section, you will learn how to install and use NetAnim to visualize the simulation execution for monitoring, debugging, and analyzing activities.

# How to install and use NetAnim

NetAnim's **Graphical User Interface (GUI)** is implemented using Qt, a cross-platform software. In order to install the NetAnim GUI, execute the following commands:

```
cd ns-allinone-3.36/netanim-3.108
make clean
qmake NetAnim.pro
make
```

After executing the preceding commands, a NetAnim executable file is going to be created in the same directory. Then, you can open it by typing ./NetAnim. This will open NetAnim's GUI (refer to *Figure 1.16*). Inside the NetAnim window, we can observe the following message: **Please select an XML trace file using the file load button on the top-left corner:**

Figure 1.16 – The ns-3 NetAnim user interface

Follow these steps to visualize your simulation program execution using NetAnim:

1.  Include the [#include "ns3/netanim-module.h"] header in your simulation program:

    A.  Include the following code: AnimationInterface anim ("animation.xml"); before Simulator::Run() in your simulation program.

    B.  Execute the program using ./ns3 run scratch/your simulation program. After successful execution, it generates an animation.xml file.

2.  Now, you can open the animation.xml trace file using NetAnim to visualize your simulation.

Next, we explore NetAnim's various options for visualizing ns-3 simulations.

## How to use the NetAnim GUI (panels and their menus)

In the NetAnim GUI, under the **Animator** tab, we can see a folder symbol. After clicking on it, select and open the `animation.xml` file. Then, it displays various nodes involved in your simulation. Now, you are ready to execute the simulation and visualize it using NetAnim by clicking on the play button from the main top panel menu.

### The main top panel menu

In the main top panel menu (refer to *Figure 1.17*), from left to right, you'll find the following options:

- The **Folder** option to load the animation XML file.

- The **Reload** option to restart the simulation.

- The **Play** button to start the simulation execution. On pressing this button, you can observe the change in the simulation time display, progress bar, packets exchange arrows, placement of nodes, and so on.

- By changing the speed slide bar, we can control the simulation execution speed.

- You can change the **Node Size** and **Lines** values to modify the displayed grid size and node sizes. Click on **MAC** or **IP** to display nodes' MAC/IP details:

Figure 1.17 – The NetAnim top panel menu items

### The side panel menu

In the side panel menu (refer to *Figure 1.18*), we can find the following important options:

Figure 1.18 – The NetAnim side panel menu items

- **Step through simulation >>**: If you want to see simulation execution step by step in simulation time, click the >> button. It allows you to see intermediate events that happened during the simulation execution. Reset the simulation by clicking on **R**.

- **Stop icon**: To enable or disable the display of packet exchange arrows.

- **Show packet metadata icon (M)**: By enabling this, we can view packets and their internal details.

- To view nodes and their details clearly, you can use the zoom-in and zoom-out options.

- **Show properties tree icon**: Clicking on this opens a complete properties description panel for nodes (refer to *Figure 1.19*):

Figure 1.19 – Showing the complete description of Node-0 in our simulation

- **Other icons**: There is an option to toggle the display of SYS node IDs and another option to display the *X* and *Y* coordinates of the mouse pointer.

## Visualizing a network simulation using NetAnim

What do you want to learn and achieve in your first network simulation visualization? Let's look at some suggestions:

- How to set up the following simple **carrier-sense multiple access** (**CSMA**) network (as shown in the comments), which connects four hosts. In ns-3, the CSMA module helps you simulate a simple shared bus network or Ethernet IEEE 802.3 and provides only basic functionality:

```
// Simple CSMA Network
//          192.168.1.0
// Node-0 ---- Node-1----Node-2----Node-3
//     CSMA Channel
// (Speed: 5Mbps, Propagation Delay: 6560 nano seconds)
// Install UDP Echo Client on Node-0
// Install UDP Echo Server on Node-3
```

- Install the UDP echo client and server applications on Node-0 and Node-1 respectively and configure the following UDP echo client application traffic characteristics:

  - MaxPackets: A maximum of 1,000 packets to be sent.

  - Interval: The speed at which the echo client application should send packets. For example, setting Interval to 0.1 means 10 packets are sent per second.

  - PacketSize: 1,024 bytes set for each packet.

- We want to monitor how the UDP echo client and server exchange packets in the network.

- Generate an animation XML file that can be used later with NetAnim. We want to visualize routing tables, packet counter values, and flow statistics using NetAnim.

- View flow-level metrics such as throughput, delay, jitter, and packet loss.

Now, we are going to explain step-by-step instructions for writing the simulation program, which will be saved under pkt_first_anim.cc in the ns-allinone-3.36/ns-3.36/scratch/ folder.

In this section, we will simulate a four-node CSMA LAN and visualize its simulation:

1.  First, import all necessary ns-3 packages (core, network, internet, applications, and flow-monitor). Besides this, as we want to simulate a CSMA LAN and visualize its execution, the csma and netanim modules are imported:

```
#include "ns3/csma-module.h"
#include "ns3/netanim-module.h"
```

2.  In ns-3, the actual simulation execution starts with the `main()` function. We write the following relevant code in `main()`:

    A.  As per the topology diagram given in our simulation activity, we start with first creating four hosts as four nodes using `NodeContainer`:

        ```
        NodeContainer nodes;
        nodes.Create (4);
        ```

    B.  Then, we create and configure the CSMA channel as per our requirements (5 Mbps speed, 6,560 ns delay) using `CsmaHelper`:

        ```
        CsmaHelper csma;
        csma.SetChannelAttribute ("DataRate", StringValue
        ("5Mbps"));
        csma.SetChannelAttribute ("Delay", TimeValue
        (NanoSeconds (6560)));
        ```

    C.  Next, install the CSMA protocol on the nodes, which connects them using the CSMA channel. As a result, it returns four devices to `NetDeviceContainer devices`: the first device is the Node-0 NIC, the second device is the Node-1 NIC, and so on:

        ```
        NetDeviceContainer devices;
        devices = csma.Install (nodes);
        ```

    D.  Now, we install TCP/IP on the nodes. Then, we configure the IP addresses for the `NetDeviceContainer` devices under the `192.168.1.0` network ID. As a result, Node-0 gets `192.168.1.1` and Node-3 gets `192.168.1.4`:

        ```
        InternetStackHelper stack;
        stack.Install (nodes);
        Ipv4AddressHelper address;
        address.SetBase ("192.168.1.0", "255.255.255.0");
        Ipv4InterfaceContainer interfaces = address.Assign
        (devices);
        ```

3.  Create the ns-3 traffic applications and install them as per the simulation topology:

    A.  First, create a UDP echo server application and install it on Node-3. Then, configure its start and stop timings:

        ```
        uint64_t port = 9;
        UdpEchoServerHelper echoServer (port);
        ApplicationContainer serverApps = echoServer.Install
        ```

```
(nodes.Get (3));
  serverApps.Start (Seconds (1.0));
  serverApps.Stop (Seconds (10.0));
```

B.   Next, create a UDP echo client application and install it on Node-0. Configure its traffic characteristics (`MaxPackets=1000`, `Interval=0.1`, and `PacketSize=1024`), and set start and stop timings based on `MaxPackets`:

```
UdpEchoClientHelper echoClient (interfaces.GetAddress
(3), port);
  echoClient.SetAttribute ("MaxPackets", UintegerValue
(1000));
  echoClient.SetAttribute ("Interval", TimeValue (Seconds
(0.1)));
  echoClient.SetAttribute ("PacketSize", UintegerValue
(1024));
  ApplicationContainer clientApps = echoClient.Install
(nodes.Get (0));
  clientApps.Start (Seconds (2.0));
  clientApps.Stop (Seconds (10.0));
```

4.   Install `FlowMonitor` on all nodes to collect flow-level statistics. Hence, at the end of the simulation, we can view each traffic flow's throughput, delay, jitter, and packet loss:

```
Ptr<FlowMonitor> flowmon;
FlowMonitorHelper flowmonHelper;
flowmon = flowmonHelper.InstallAll ();
```

5.   Configure the animation interface. At the end of the simulation execution, an animation XML file called `first_animation_demo.xml` will be generated. The XML file can be viewed using NetAnim to visualize the simulation execution:

A.   We want to name Node-0 as a UDP echo client and Node-3 as a UDP echo server in the visualization:

```
AnimationInterface anim ("first_animation_demo.xml");
anim.UpdateNodeDescription (3, "UDP Echo Server");
anim.UpdateNodeDescription (0, "UDP Echo Client");
```

B.   During the simulation execution, we want to visualize the various protocol packets exchanged:

```
anim.EnablePacketMetadata ();
```

C.  At end of the simulation, we want to visualize routing tables configured on various nodes using NetAnim:

```
anim.EnableIpv4RouteTracking ("firstanim_routetable.
xml", Seconds (0), Seconds (2), Seconds (0.25));
```

D.  At end of the simulation, we want to visualize various packet counter values of various nodes using NetAnim:

```
anim.EnableIpv4L3ProtocolCounters (Seconds (0), Seconds
(10));
anim.EnableQueueCounters (Seconds (0), Seconds (10));
Simulator::Stop (Seconds(11.0));
Simulator::Run ();
```

E.  At end of the simulation, we want to view all flows' flow statistics, such as throughput, delay, jitter, and packet loss:

```
flowmon->SerializeToXmlFile ("firstanim_flowmetrics.
xml", true, true);
Simulator::Destroy ();
```

Now, you can execute your simulation program using the following command:

```
$ ./ns3 run scratch/pkt_first_anim
```

It generates 3 XML files such as first_animation_demo.xml, firstanim_routetable.xml, and firstanim_flowmetrics.xml

## Playing your simulation using NetAnim

Now, play your simulation using NetAnim and do the following - After opening NetAnim, open `first_animation_demo.xml` to visualize the simulation. While playing this simulation using NetAnim, you can see (refer to *Figure 1.20*) the first exchange of **Address Resolution Protocol** (**ARP**) packets to discover the link addresses (**MAC**) of the nodes, then the actual UDP traffic packet exchange begins among the nodes:

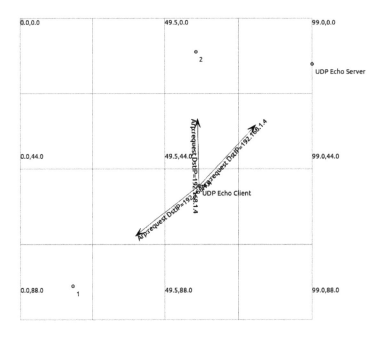

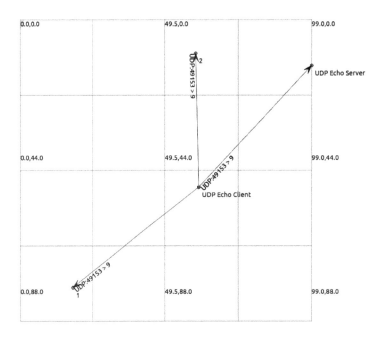

Figure 1.20 – Visualize ARP packets and UDP packets using NetAnim

> **Important note**
>
> If you comment `anim.EnablePacketMetadata()` in the code or disable the **M** option from the side panel, you will not see any packets exchanged during the simulation visualization. At the end of the simulation, if you want to see packet exchanges happening between various nodes, you can do it using the **Packets** tab options. We'll explore the **Packets** tab and its options in the following subsection.

### Inspecting a variety of traffic packets under the NetAnim Packets tab

Under the **Packets** tab (refer to *Figure 1.21*), by using the **Show Nodes** option and selecting the packet type, it is possible to view various packets exchanged between various nodes:

Figure 1.21 – The NetAnim Packets tab options

For example, we can observe the exchange of ARP packets between Node-0 and Node-3 by doing the following (refer to *Figure 1.22*). Type 0  :  :  : 3 in **Show Nodes**, tick **Arp**, then click on **Submit**:

Figure 1.22 – Select the Arp option to view ARP packets exchange

We can observe the following chart (refer to *Figure 1.23*) showing the exchange of ARP packets between Node-0 and Node-3:

Figure 1.23 – ARP packets exchange between Node-0 and Node-3 shown as a sequence diagram

We can also observe the exchange of ARP packets between Node-0 and Node-3 in a table format (refer to *Figure 1.24*) in time series rows:

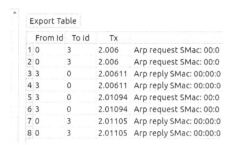

Figure 1.24 – ARP packets exchange shown as a table

Similarly, you can try observing Ethernet, IP, and UDP packet exchanges between various nodes at simulation time. Next, we will explore the **Stats** tab for visualizing routing tables, various counter values, and flow statistics.

### Inspecting simulation statistics under the NetAnim Stats tab

In order to inspect network node configuration details such as IP and MAC, flow-level stats, routing tables, and counters, we can use the **Stats** tab (refer to *Figure 1.25*) and its following options:

Figure 1.25 – The NetAnim Stats tab options

The first option is **IP-MAC**. By default, IP-MAC details (refer to *Figure 1.26*) can be viewed without any lines of code in your simulation program:

Figure 1.26 – The NetAnim IP-MAC menu window

As we selected the **All** option, it is displaying all the nodes' IP and MAC details, as shown in *Figure 1.27*. You can try selecting a particular node ID:

Figure 1.27 – The ns-3 nodes' IP and MAC details

We also have the **Routing** option. To view routing tables, open `firstanim_routetable.xml`, which was generated by your simulation program. Now, we can view the routing tables of all nodes or individual nodes by clicking on the node numbers displayed on the side panel shown in *Figure 1.28*:

Figure 1.28 – The NetAnim Routing menu window

We selected the **All** option to display all nodes' routing tables, as shown in *Figure 1.29*:

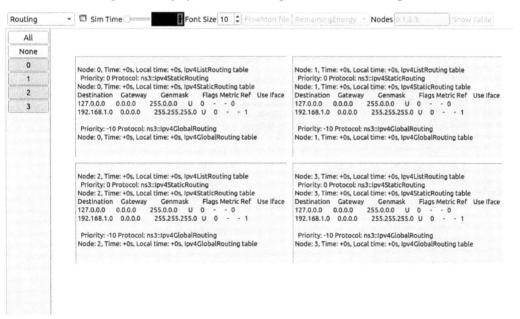

Figure 1.29 – Our simulation ns-3 nodes' routing table details

Then, we have the **Flow-monitor** option. From the **Stats** tab, select **Flow-monitor** (refer to *Figure 1.30*). Next, click the **FlowMon file** button and open `firstanim_flowmetrics.xml`, generated by your simulation program:

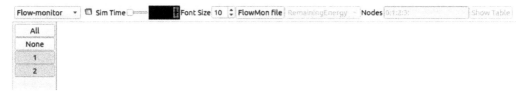

Figure 1.30 – The NetAnim Flow-monitor window

Now, we can view the flow-level stats of all nodes (refer to *Figure 1.31*) or individual nodes by clicking on the node numbers displayed on the side panel. We selected the **ALL** option to show all flows' statistics:

Figure 1.31 – Our simulation flow-level statistics

We also have the **Counter Tables** option. We can check the counter value of various nodes by selecting the **Counter Tables** menu item under the **Stats** menu list.

After selecting **Counter Tables**, we can select from the **Ipv4 Tx**, **Ipv4 Rx**, **Ipv4 Drop**, **Enqueue**, **Dequeue**, and **Queue Drop** options next to the **FlowMon file** option (refer to *Figure 1.32*):

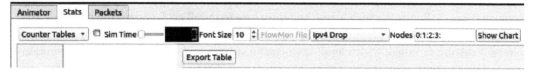

Figure 1.32 – The NetAnim Counter Tables window

For example, to show IPv4 packet drops on various nodes, you can select the **Ipv4 Drop** option and view the result (refer to *Figure 1.33*) in table or chart format by clicking on the **show table** or **show chart** button:

| | Time | 0 | 1 | 2 | 3 |
|---|---|---|---|---|---|
| 1 | 0 | 0 | | | |
| 2 | 0 | | 0 | | |
| 3 | 0 | | | 0 | |
| 4 | 0 | | | | 0 |
| 5 | 0 | 0 | | | |
| 6 | 0 | | 0 | | |
| 7 | 0 | | | 0 | |
| 8 | 0 | | | | 0 |
| 9 | 1 | 0 | | | |
| 10 | 1 | | 0 | | |
| 11 | 1 | | | 0 | |
| 12 | 1 | | | | 0 |
| 13 | 2 | 0 | | | |
| 14 | 2 | | 0 | | |
| 15 | 2 | | | 0 | |
| 16 | 2 | | | | 0 |
| 17 | 3 | 0 | | | |
| 18 | 3 | | 0 | | |
| 19 | 3 | | | 0 | |
| 20 | 3 | | | | 0 |
| 21 | 4 | 0 | | | |
| 22 | 4 | | 0 | | |
| 23 | 4 | | | 0 | |

Figure 1.33 – Our simulation nodes' IPv4 packet drop counter table

Well done! After practicing your first hands-on activity, you are ready to explore ns-3's more interesting features.

# Summary

Congratulations on the successful completion of this chapter. We have just completed this starting session on the ns-3 simulator, which helped you learn how to install ns-3, understand ns-3 code and simulation structures, and write simulation programs using the `Code::Blocks` editor. We also explored NetAnim thoroughly with its all-important features for visualizing simulation execution results and details. Next, we are going to start another exciting chapter that discusses important features of ns-3, such as logging, debugging, tracing, and how to validate simulation results.

# 2

# Monitoring, Debugging, Tracing, and Evaluating Network Topologies in ns-3

In order to set up and evaluate complex simulations, it is necessary to learn how to display various simulation details, such as information, events, and errors, as well as handling errors, crashes, and debugging. It also involves inspecting nodes and protocols of simulation topologies. In this chapter, to carry out the aforementioned tasks to perfection, we discuss the important features of ns-3, such as logging, debugging, tracing, and validating simulation results.

Logging is an important ns-3 feature that helps users to monitor simulation details such as events, logic, warnings, errors, and debug messages. ns-3 supports a flexible way to include log statements and enable or disable logging details in simulation programs. Another important feature ns-3 users must know of is debugging. It is common to face various errors and crashes during simulation execution. In order to resolve simulation errors and crashes quickly, similar to any C++ program debugging, ns-3 users can also use gdb.

The primary objective of running simulation programs is to analyze the working of a variety of nodes and their protocols and algorithms. In ns-3, by using the tracing feature, users can perform a careful inspection of nodes and their protocols and algorithms. For researchers or engineers to validate their simulations, it is necessary to compute important performance metrics, such as throughput, delay, and packet loss, and project those metrics visually. ns-3 users can use the **Flow Monitor** module to compute flow-level performance metrics based on flow type. At the end of this chapter, we also discuss how to use Flow Monitor results and project the performance metrics using a variety of **GNU** plots visually. The following topics are discussed in this chapter:

- Monitoring ns-3 simulation steps and events using logging

- Identifying and resolving simulation issues using a debugger

- Using callbacks and traces for inspection of ns-3 nodes and protocols

- Performance evaluation and validation of simulations using `FlowMonitor`
- Using `gnuplot` to project simulation results

# Monitoring ns-3 simulation steps and events using logging

ns-3 supports logging features for all its supported modules. Logging enables users to view various important details, such as the working of a protocol or an application, while executing their simulation. The ns-3 developers have included lots of useful logging statements in their modules, and ns-3 users can enable the logging details of the modules they're interested in in their simulation programs. In order to provide finer control over which logging statements are to be displayed during simulation execution, ns-3 supports the following levels of logging statements:

- `LOG_INFO`
- `LOG_DEBUG`
- `LOG_FUNCTION`
- `LOG_WARN`
- `LOG_ERROR`
- `LOG_ASSERT`
- `LOG_FATAL_ERROR`
- `LOG_ALL`

ns-3 users can select any level and display details in their simulations. Users can display all levels of logging statements of a particular module, by selecting the `LOG_ALL` level for a particular module in their simulation program. However, it may be clumsy in terms of simulation details display. Hence, it is suggested to enable only the necessary levels to display.

We discuss ns-3 logging features in two ways:

- How to display existing modules' logging details at various levels
- How to include your own logging statements in your simulation program

## How to display existing modules' logging details at various levels

Let's see how to include and display existing ns-3 modules' logging statements in your simulation. For example, to include log components of UDP echo client and server applications in the `pkt_first.cc` simulation, we carry out the following procedure.

First, find out all the log statements that are offered by `UdpEchoClientApplication` and `UdpEchoServerApplication`. To do this, we can go to `ns-allinone-3.36/ns-3.36/src/applications/model` and enter the following commands:

```
grep 'LOG' udp-echo-client.cc
```

It displays the following matching logging statements of the UDP client:

```
NS_LOG_FUNCTION (this);
    NS_LOG_INFO ("At time " << Simulator::Now ().As (Time::S) << " client sent " << m_size << " bytes to " <<
    NS_LOG_INFO ("At time " << Simulator::Now ().As (Time::S) << " client sent " << m_size << " bytes to " <<
    NS_LOG_INFO ("At time " << Simulator::Now ().As (Time::S) << " client sent " << m_size << " bytes to " <<
    NS_LOG_INFO ("At time " << Simulator::Now ().As (Time::S) << " client sent " << m_size << " bytes to " <<
NS_LOG_FUNCTION (this << socket);
        NS_LOG_INFO ("At time " << Simulator::Now ().As (Time::S) << " client received " << packet->GetSize () << " bytes from " <<
        NS_LOG_INFO ("At time " << Simulator::Now ().As (Time::S) << " client received " << packet->GetSize () << " bytes from " <<
```

Figure 2.1 – The UDP client log statements

```
grep 'LOG' udp-echo-server.cc
```

It displays the following matching logging statements of the UDP server:

```
NS_LOG_FUNCTION (this << socket);
        NS_LOG_INFO ("At time " << Simulator::Now ().As (Time::S) << " server received " << packet->GetSize () << " bytes from " <<
        NS_LOG_INFO ("At time " << Simulator::Now ().As (Time::S) << " server received " << packet->GetSize () << " bytes from " <<
    NS_LOG_LOGIC ("Echoing packet");
        NS_LOG_INFO ("At time " << Simulator::Now ().As (Time::S) << " server sent " << packet->GetSize () << " bytes to " <<
        NS_LOG_INFO ("At time " << Simulator::Now ().As (Time::S) << " server sent " << packet->GetSize () << " bytes to " <<
```

Figure 2.2 – The UDP server log statements

The `NS_LOG_FUN` and `NS_LOG_INFO` statements are available in UDP client/server programs. We can enable the `LOG_INFO` statements in `pkt_first.cc` to observe simulation details such as when the client/server sent packets and when the client/server received packets.

Next, let's see how to include logging statements of the UDP echo client or server application in a simulation program. To include logging statements of these applications, we need to check the `LogComponent` names in their `.cc` files (e.g., `NS_LOG_COMPONENT_DEFINE ("UdpEchoClientApplication")` statements). Then, we can enable the log component of the module we're interested in our simulation program by including the following lines of code in `pkt_first.cc`:

```
LogComponentEnable ("UdpEchoClientApplication", LOG_LEVEL_
INFO);
LogComponentEnable ("UdpEchoServerApplication", LOG_LEVEL_
INFO);
```

In the `LogComponentEnable` statement, the first argument is the *log component name* and the second argument is the *logging level* to display. We can include the level we are interested in by enabling the log component in the `pkt_first.cc` simulation program, and executing it. By enabling `LOG_INFO`, we can observe the following simulation output:

```
At time +2s client sent 1024 bytes to 192.168.1.2 port 9
At time +2.00101s server received 1024 bytes from 192.168.1.1 port 49153
At time +2.00101s server sent 1024 bytes to 192.168.1.1 port 49153
At time +2.00202s client received 1024 bytes from 192.168.1.2 port 9
At time +2.1s client sent 1024 bytes to 192.168.1.2 port 9
At time +2.10101s server received 1024 bytes from 192.168.1.1 port 49153
At time +2.10101s server sent 1024 bytes to 192.168.1.1 port 49153
At time +2.10202s client received 1024 bytes from 192.168.1.2 port 9
At time +2.2s client sent 1024 bytes to 192.168.1.2 port 9
At time +2.20101s server received 1024 bytes from 192.168.1.1 port 49153
At time +2.20101s server sent 1024 bytes to 192.168.1.1 port 49153
At time +2.20202s client received 1024 bytes from 192.168.1.2 port 9
At time +2.3s client sent 1024 bytes to 192.168.1.2 port 9
At time +2.30101s server received 1024 bytes from 192.168.1.1 port 49153
At time +2.30101s server sent 1024 bytes to 192.168.1.1 port 49153
At time +2.30202s client received 1024 bytes from 192.168.1.2 port 9
At time +2.4s client sent 1024 bytes to 192.168.1.2 port 9
```

Figure 2.3 – The UDP client and server LOG_INFO statements

That's simple. Next, we see how to include our simulation log statements and test them.

## How to include your own logging statements in your simulation program

We will learn how to do this activity by defining a `FirstLoggingExample` log component in our `pkt_first_log.cc`. In `FirstLoggingExample`, various log-level statements are included and tested as follows:

- In `pkt_first_log.cc`, we create 100 nodes and connect them using a CSMA channel. We also install an internet stack and configure IP addresses.

- While creating nodes, we want to display successful node creation information.

- Before connecting nodes using the CSMA channel, we want to warn the user to include `csma-module`.

- For configuring IP addresses to nodes, we need to instruct users first to install the TCP/IP stack and then to configure IP addresses in a logical order.

- Use a debugging log statement for indicating `NetDevice` indices details.

- If the wrong index is passed, we want to display respective error log statements.

Now, let's write the necessary code in pkt_first_log.cc and test it:

1.  Include all necessary ns-3 header files:

    ```
    #include "ns3/core-module.h"
    #include "ns3/network-module.h"
    #include "ns3/internet-module.h"
    #include "ns3/csma-module.h"
    using namespace ns3;
    ```

2.  Define the log component for your simulation as FirstLoggingExample:

    ```
    NS_LOG_COMPONENT_DEFINE ("FirstLoggingExample");
    int
    main (int argc, char *argv[])
    {
      NS_LOG_FUNCTION (argc);
    ```

3.  Write information-level log statements related to the node creation status:

    ```
    NS_LOG_INFO ("We are going to simulate a network with
    100 Nodes");
      NodeContainer nodes;
      nodes.Create (100);;
      NS_LOG_INFO ("100 Nodes are created successfully");
    ```

4.  Write warning-level log statements related to connecting nodes using the CSMA channel:

    ```
    NS_LOG_WARN ("Must import csma-module to use it");
    CsmaHelper csma;
    NetDeviceContainer devices = csma.Install(nodes);
    ```

5.  Write logic-level log statements related to installing the internet stack and configuring IP addresses:

    ```
    NS_LOG_LOGIC ("First Install TCP/IP Internet Stack");
    InternetStackHelper stack;
    stack.Install (nodes);
    NS_LOG_LOGIC ("After installing TCP/IP Internet Stack
    you can assign IP address to Nodes");
      Ipv4AddressHelper address;
      address.SetBase ("10.1.1.0", "255.255.255.0");
    ```

```
Ipv4InterfaceContainer interfaces = address.Assign
(devices);
```

6.  Write debug-level log statements related to accessing net devices:

```
NS_LOG_DEBUG("NetDevices  index start with Zero to
print Nth node address give index as N-1"<<interfaces.
GetAddress (1));
int index=100;
```

7.  Write error-level log statements related to accessing net devices using suitable indices:

```
if (index>99)
NS_LOG_ERROR ("No device is available");
index=1;
NS_ASSERT_MSG (index>=0, "index must be > 0 invalid
index.");
Simulator::Run ();
Simulator::Destroy ();
}
```

We have included various log-level statements for our pkt_first_log.cc simulation program. As FirstLoggingExample is a new logging module, we need to execute the following commands from the ns-allinone-3.36/ns-3.36 folder to include it in the ns-3 log components.

First, include FirstLoggingExample in the ns-3 log components list by building ns-3 using the following command:

```
./ns3
```

Clear all existing exported log components list and run your simulation:

```
export NS_LOG=
./ns3 run scratch/pkt_first_log
```

Next, export FirstLoggingExample and enable all log levels:

```
export NS_LOG=FirstLoggingExample=all
./ns3 run scratch/pkt_first_log
```

Finally, we can see all level log statements of the `FirstLoggingExample` component in *Figure 2.4*:

```
FirstLoggingExample:main(1)
We are going to simulate a network with 100 Nodes
100 Nodes are created successfully
Must import csma-module to use it
First Install TCP/IP Internet Stack
After installing TCP/IP Internet Stack you can assign IP address to Nodes
NetDevices  index start with Zero to print Nth node address give index as N-110.1.1.2
No device is available
```

Figure 2.4 – All log statements of FirstLoggingExample

As it displays all log statements, it may be difficult to understand the simulation details if there are a large number of logging statements. Now, we enable only certain log levels such as LOG_INFO and LOG_LOGIC by running the following commands:

```
export NS_LOG='FirstLoggingExample=info|logic'
./ns3 run scratch/pkt_first_log
```

Only LOG_INFO or LOG_LOGIC statements are displayed in *Figure 2.5*:

```
We are going to simulate a network with 100 Nodes
100 Nodes are created successfully
First Install TCP/IP Internet Stack
After installing TCP/IP Internet Stack you can assign IP address to Nodes
```

Figure 2.5 – INFO or LOGIC log statements of FirstLoggingExample

That's great. We have learned how to use various modules' log statements as well as how to include log statements in our simulations. Next, we will learn how to handle various simulation issues using the gdb standard debugging tool.

# Identifying and resolving simulation issues using a debugger

It is very hard to run a simulation without facing any errors and crashes. Besides, running a simulation also involves changing and tuning the application's and protocol's parameters to resolve various issues. As end users, we expect issues to be identified and resolved quickly. For instance, say a simulation aborted or crashed suddenly without giving any details. In order to resolve this particular issue, we may try to go through the simulation script or code and use log statements to identify the issue. This process is tedious and it takes longer to identify the actual reason and lines of code to change. Hence, users look for quick and easy debugging solutions, for example, using debuggers, such as gdb. ns-3 also supports the use of gdb for debugging simulation programs just like any C++ program. In this section, we discuss how to use gdb for debugging simulations. We will go through the following:

- Going through the gdb commands quickly

- Walking through a simulation using gdb

- Detecting and resolving a simulation crash using gdb

## Going through the gdb commands quickly

We suggest you go through the following necessary gdb commands and their usage for debugging simulation programs:

| Usage | gdb command |
| --- | --- |
| Insert a breakpoint | break or b [file: ] linenumber. Filename is optional |
| Enable or disable a breakpoint | enable breakpoint number or disable breakpoint number |
| Display all breakpoints' details | info break |
| See lines of code to insert breakpoint | list |
| Execute the program using gdb | r |
| Inspect a function's code | step or s |
| Line-by-line execution using gdb | next or n |
| Inspect a callback trace | bt |
| Skip a few lines of code | until  linenumber |
| Inspect local arguments | Info args |
| Inspect function arguments | Info locals |
| Print variable values | print or p variable name |
| Set variable value | set var varname=value |

Next, we will do a hands-on task related to how to use gdb in ns-3.

## Walking through a simulation using gdb

In this hands-on debugging activity, our objective is to set two breakpoints in pkt_first.cc, one for viewing the node's Create function lines of code, and another breakpoint to change MaxPackets of the UDP client. Let's first set two breakpoints at 51 and 75 in pkt_first.cc:

```
(gdb) b 51
```

This sets a breakpoint at line number `51` (refer to *Figure 2.6*):

```
(gdb) b 51
Breakpoint 1 at 0x55de: file /home/iiitdmk/ns-allinone-3.36/ns-3.36/scratch/pkt_first.cc, line 51.
(gdb) list
52
53          PointToPointHelper pointToPoint;
54          pointToPoint.SetDeviceAttribute ("DataRate", StringValue ("1Gbps"));
55          pointToPoint.SetChannelAttribute ("Delay", StringValue ("1ms"));
```

Figure 2.6 – Setting a breakpoint at line number 51 in pkt_first.cc

After setting breakpoints at line numbers `51` and `75`, execute the simulation by running the `r` command. Then, execution stops at line `51` (shown in *Figure 2.7*):

```
(gdb)  b  75

(gdb)  r
```

```
(gdb) b 75
Breakpoint 2 at 0x5987: file /home/iiitdmk/ns-allinone-3.36/ns-3.36/scratch/pkt_first.cc, line 75.
(gdb) r
Starting program: /home/iiitdmk/ns-allinone-3.36/ns-3.36/build/scratch/ns3.36-pkt_first-default
[Thread debugging using libthread_db enabled]
Using host libthread_db library "/lib/x86_64-linux-gnu/libthread_db.so.1".

Breakpoint 1, main (argc=<optimized out>, argv=<optimized out>) at /home/iiitdmk/ns-allinone-3.36/ns-3.36/scratch/pkt_first.cc:51
51          nodes.Create (2);
```

Figure 2.7 – Result of executing the r gdb command after setting breakpoints

Next, step into the `Create` function using the `s` command to go through its lines of code using the `n` command. Repeatedly execute the `n` command till the main program has been reached (shown in *Figure 2.8*):

```
(gdb) s
ns3::NodeContainer::Create (this=0x7fffffffd390, n=2) at /home/iiitdmk/ns-allinone-3.36/ns-3.36/src/network/helper/node-container.cc:99
99      {
(gdb) n
100         for (uint32_t i = 0; i < n; i++)
(gdb) n
102             m_nodes.push_back (CreateObject<Node> ());
(gdb) n
100         for (uint32_t i = 0; i < n; i++)
(gdb) n
102             m_nodes.push_back (CreateObject<Node> ());
(gdb) n
100         for (uint32_t i = 0; i < n; i++)
(gdb) n
main (argc=<optimized out>, argv=<optimized out>) at /home/iiitdmk/ns-allinone-3.36/ns-3.36/scratch/pkt_first.cc:53
53          PointToPointHelper pointToPoint;
```

Figure 2.8 – Stepping into the Create function and inspecting the code

As there are many lines of code, let's check the list of breakpoints available in our simulation and skip through a few lines of code using the `info` command:

```
(gdb)  info  b
```

This displays all breakpoints of your simulation program (shown in *Figure 2.9*):

```
(gdb) info b
Num     Type           Disp Enb Address            What
1       breakpoint     keep y   0x00805555555595de in main(int, char**) at /home/iiitdmk/ns-allinone-3.36/ns-3.36/scratch/pkt_first.cc:51
        breakpoint already hit 1 time
2       breakpoint     keep y   0x0080555555559907 in main(int, char**) at /home/iiitdmk/ns-allinone-3.36/ns-3.36/scratch/pkt_first.cc:75
```

Figure 2.9 – List of breakpoints and their numbers in your simulation program

As we see many lines of code between 51 and 75, we want to resume stepping through code again from line 70 using the until 70 command. It stops simulation execution at line 70. Then, we can step through the simulation again using the n command till we reach breakpoint 2 to set MaxPackets:

**(gdb) until 70**

This stops debugging at line number 70 of your simulation program (shown in *Figure 2.10*):

```
(gdb) until 70
main (argc=<optimized out>, argv=<optimized out>) at /home/iiitdmk/ns-allinone-3.36/ns-3.36/scratch/pkt_first.cc:70
70          ApplicationContainer serverApps = echoServer.Install (nodes.Get (1));
(gdb) n
71          serverApps.Start (Seconds (1.0));
(gdb) n
72          serverApps.Stop (Seconds (10.0));
(gdb) n
74          UdpEchoClientHelper echoClient (interfaces.GetAddress (1), port);
(gdb) n

Breakpoint 2, main (argc=<optimized out>, argv=<optimized out>) at /home/iiitdmk/ns-allinone-3.36/ns-3.36/scratch/pkt_first.cc:75
75          echoClient.SetAttribute ("MaxPackets", UIntegerValue (10));
```

Figure 2.10 – until command usage during debugging

Next, at breakpoint 2, we want to step into SetAttribute and change MaxPackets to 4 using the following command:

**(gdb) s**

After stepping into SetAttribute(), then step through the main simulation execution using n for a few lines of code. Then, finally, run the c command to continue the execution (shown in *Figure 2.11*):

**(gdb) c**

```
(gdb) set var value=4
(gdb) n
main (argc=<optimized out>, argv=<optimized out>) at /usr/include/c++/9/ext/new_allocator.h:80
80              new_allocator() _GLIBCXX_USE_NOEXCEPT { }
(gdb) n
75          echoClient.SetAttribute ("MaxPackets", UIntegerValue (10));
(gdb) n
76          echoClient.SetAttribute ("Interval", TimeValue (Seconds (0.1)));
(gdb) n
77          echoClient.SetAttribute ("PacketSize", UIntegerValue (1024));
(gdb) n
80              new_allocator() _GLIBCXX_USE_NOEXCEPT { }
(gdb) n
77          echoClient.SetAttribute ("PacketSize", UIntegerValue (1024));
(gdb) n
79          ApplicationContainer clientApps = echoClient.Install (nodes.Get (0));
```

Figure 2.11 – Changing MaxPackets during debugging

We can see that instead of 10 packets, only 4 packets are exchanged between the client and server in *Figure 2.12*:

```
At time +2s client sent 1024 bytes to 192.168.1.2 port 9
At time +2.00101s server received 1024 bytes from 192.168.1.1 port 49153
At time +2.00101s server sent 1024 bytes to 192.168.1.1 port 49153
At time +2.00202s client received 1024 bytes from 192.168.1.2 port 9
At time +2.1s client sent 1024 bytes to 192.168.1.2 port 9
At time +2.10101s server received 1024 bytes from 192.168.1.1 port 49153
At time +2.10101s server sent 1024 bytes to 192.168.1.1 port 49153
At time +2.10202s client received 1024 bytes from 192.168.1.2 port 9
At time +2.2s client sent 1024 bytes to 192.168.1.2 port 9
At time +2.20101s server received 1024 bytes from 192.168.1.1 port 49153
At time +2.20101s server sent 1024 bytes to 192.168.1.1 port 49153
At time +2.20202s client received 1024 bytes from 192.168.1.2 port 9
At time +2.3s client sent 1024 bytes to 192.168.1.2 port 9
At time +2.30101s server received 1024 bytes from 192.168.1.1 port 49153
At time +2.30101s server sent 1024 bytes to 192.168.1.1 port 49153
At time +2.30202s client received 1024 bytes from 192.168.1.2 port 9
[Inferior 1 (process 12666) exited normally]
(gdb)
```

Figure 2.12 – Simulation results after changing MaxPackets to 4

As a hands-on task, users can try changing other UDP echo client traffic parameters too, and observe results while debugging pkt_first.cc. That's great. We have successfully completed a gdb session. Next, we'll see how to handle a simulation crash using gdb.

## Detecting and resolving a simulation crash

We changed pkt_first.cc intentionally to simulate a crash. We will identify and resolve the crash using gdb. Let's execute our simulation using the following command:

```
$./ns3 run scratch/pkt_first
```

After executing the command, we can observe our simulation crashes (shown in *Figure 2.13*):

```
Scanning dependencies of target scratch_pkt_first
[  0%] Building CXX object scratch/CMakeFiles/scratch_pkt_first.dir/pkt_first.cc.o
[  0%] Linking CXX executable ../../build/scratch/ns3.36-pkt_first-default
Command 'build/scratch/ns3.36-pkt_first-default' died with <Signals.SIGSEGV: 11>.
```

Figure 2.13 – Simulation crash results

Oops! Our simulation crashes. There is no clue given as to what the error is. How can we resolve this crash? Going through our simulation code, we need to enable logging, and inspect many lines of code. As we have a large number of lines of code, this process is not feasible to identify and resolve the issue quickly. Then, can we use debugging? Yes, we can use gdb to get details related to the crash. It helps you identify specific lines of code to inspect quickly. We can look into the lines of code and resolve the issue.

Just run our simulation in debugging mode, as shown in *Figure 2.14*:

```
(gdb) r
Starting program: /home/iiitdmk/ns-allinone-3.36/ns-3.36/build/scratch/ns3.36-pkt_first-default
[Thread debugging using libthread_db enabled]
Using host libthread_db library "/lib/x86_64-linux-gnu/libthread_db.so.1".

Program received signal SIGSEGV, Segmentation fault.
ns3::Ptr<ns3::Node>::Acquire (this=0x7fffffffd4b0) at /home/iiitdmk/ns-allinone-3.36/ns-3.36/build/include/ns3/ptr.h:579
579             m_ptr->Ref ();
```

Figure 2.14 – Simulation crash details displayed using gdb

It is showing that the simulation crashed due to a segmentation fault. But, still, we do not know which lines of our simulation code are causing the crash. To get more details, type the following gdb command: backtrace. This gives the details of function calls and their arguments leading to the crash:

```
(gdb) bt
```

We can see a list of function calls leading to the crash in *Figure 2.15*:

```
#0  ns3::Ptr<ns3::Node>::Acquire (this=0x7fffffffd4b0) at /home/iiitdmk/ns-allinone-3.36/ns-3.36/build/include/ns3/ptr.h:579
#1  ns3::Ptr<ns3::Node>::Ptr (o=..., this=0x7fffffffd4b0) at /home/iiitdmk/ns-allinone-3.36/ns-3.36/build/include/ns3/ptr.h:609
#2  ns3::NodeContainer::Get (this=0x7fffffffd390, i=<optimized out>) at /home/iiitdmk/ns-allinone-3.36/ns-3.36/src/network/helper/node-container.cc:95
#3  0x00005555555598c3 in main (argc=<optimized out>, argv=<optimized out>) at /home/iiitdmk/ns-allinone-3.36/ns-3.36/scratch/pkt_first.cc:70
```

Figure 2.15 – Displaying our simulation execution function call stack

Now, we can observe that a line of code at line number 70 in pkt_first.cc is leading to Get | Ptr | Acquire function calls and causing a segmentation fault. By inspecting the following line of code in pkt_first.cc:line number 70, the user can solve the issue:

```
ApplicationContainer serverApps = echoServer.Install (nodes.Get
(2));
```

In our simulation, only two nodes are available, hence the issue must be due to passing the invalid index in nodes.Get(). Let's change the index to 1, as given in our topology, to resolve the crash and run the simulation normally. Quit the previous gdb session and run your simulation in debugging mode again using the following commands:

```
(gdb) q
$ /ns3 run scratch/pkt_first -gdb
```

As we know, line 70 in pkt_first.cc is causing a crash. We can try to change the line of code using the following commands:

```
(gdb) b 70
(gdb) r
```

Then, your simulation stops at line number 70, as shown in *Figure 2.16*:

```
Starting program: /home/iiitdmk/ns-allinone-3.36/ns-3.36/build/scratch/ns3.36-pkt_first-default
[Thread debugging using libthread_db enabled]
Using host libthread_db library "/lib/x86_64-linux-gnu/libthread_db.so.1".

Breakpoint 1, main (argc=<optimized out>, argv=<optimized out>) at /home/iiitdmk/ns-allinone-3.36/ns-3.36/scratch/pkt_first.cc:70
70          ApplicationContainer serverApps = echoServer.Install (nodes.Get (2));
```

Figure 2.16 – Inspecting line number 70 to see the crash details

Then, step into `echoServer.Install (nodes.Get(2))` and change the index value to `1` using the following commands:

```
(gdb) s
(gdb) set var i=1
```

Now, you can test the simulation execution by running the continue command: `c`. Then, you can see normal execution results, as in *Figure 2.17*:

```
(gdb) c
```

This displays UDP client and server packet exchange results shown in *Figure 2.17*:

```
At time +2.80101s server sent 1024 bytes to 192.168.1.1 port 49153
At time +2.80202s client received 1024 bytes from 192.168.1.2 port 9
At time +2.9s client sent 1024 bytes to 192.168.1.2 port 9
At time +2.90101s server received 1024 bytes from 192.168.1.1 port 49153
At time +2.90101s server sent 1024 bytes to 192.168.1.1 port 49153
At time +2.90202s client received 1024 bytes from 192.168.1.2 port 9
[Inferior 1 (process 5103) exited normally]
```

Figure 2.17 – Simulation results after resolving the crash using gdb

Finally, you can open `pkt_first.cc` and change the following line of code to permanently fix the issue:

```
ApplicationContainer serverApps = echoServer.Install (nodes.Get
(1));
```

Congratulations! You have learned how to use gdb to inspect the simulation code and handle various simulation errors and crashes.

In the next section, we are going to learn about another powerful feature of ns-3, called tracing. This feature helps users to quickly understand, use, and access a variety of network protocols and applications in simulations.

# Using callbacks and traces for inspection of ns-3 nodes and protocols

In simulations, it is very important to inspect the working of protocols and applications. ns-3 offers an important feature called tracing to inspect a variety of nodes and their installed protocols and applications. Before using the ns-3 tracing feature in your simulation, it is important to learn how tracing works. In ns-3 simulations, nodes are the important building blocks to understand where we install protocols and applications. In ns-3, tracing can be viewed as the implementation of event handling at sources and sinks. In ns-3, nodes are major event sources, where protocols and applications generate a variety of events, such as packets received and dropped and congestion window changed. These events carry important details related to protocols and applications from sources. In ns-3, trace sources are implemented using the ns-3 `TracedCallback` class.

On the other hand, while running the simulation, we are interested in listening to various events and gathering important details of nodes' protocols and applications. Hence, our simulation programs are *sinks* and they listen for interested events and process the details.

Now, we'll discuss the implementation details of ns-3 traces. In ns-3, trace sources are declared using the ns-3 `TracedCallback` class in various modules, where events are generated. Besides, ns-3 offers `Config::Connect` or `Config::ConnectWithoutContext` to connect the *trace sources* using callback functions. We will carry out the following hands-on activities related to ns-3 traces:

- Accessing CSMA protocol-processed packets in your simulation
- Understanding TCP protocol implementation details in ns-3

## Accessing CSMA protocol-processed packets in your simulation

This activity can be done by following two simple steps:

1. Identify interested trace sources to connect them in your simulation.
2. Write the necessary callback functions for trace sources and connect them in your simulation.

### Identifying interested trace sources to connect them in your simulation

ns-3 supports many trace sources for its modules. In order to use these trace sources in your simulation program, we suggest the following procedure. For example, in our simulation, we are interested to know at various nodes which frames are successfully transmitted, backoff, and so on. To find the corresponding trace sources, you can follow this procedure.

As we are interested in knowing the list of trace sources supported in the CSMA module, we can execute the following Unix command in the `src/csma/model` folder of our ns-3 installation path:

```
grep 'AddTraceSource' *
```

All supported trace sources of the CSMA module are displayed in *Figure 2.18*:

```
csma-net-device.cc:      .AddTraceSource ("MacTx",
csma-net-device.cc:      .AddTraceSource ("MacTxDrop",
csma-net-device.cc:      .AddTraceSource ("MacPromiscRx",
csma-net-device.cc:      .AddTraceSource ("MacRx",
csma-net-device.cc:      .AddTraceSource ("MacRxDrop",
csma-net-device.cc:      .AddTraceSource ("MacTxBackoff",
csma-net-device.cc:      .AddTraceSource ("PhyTxBegin",
csma-net-device.cc:      .AddTraceSource ("PhyTxEnd",
csma-net-device.cc:      .AddTraceSource ("PhyTxDrop",
csma-net-device.cc:      .AddTraceSource ("PhyRxBegin",
csma-net-device.cc:      .AddTraceSource ("PhyRxEnd",
csma-net-device.cc:      .AddTraceSource ("PhyRxDrop",
csma-net-device.cc:      .AddTraceSource ("Sniffer",
csma-net-device.cc:      .AddTraceSource ("PromiscSniffer",
```

Figure 2.18 – List of trace sources available in the CSMA module

From the various trace sources displayed in *Figure 2.18*, we can identify a trace source that interests us, and understand its `TracedCallback` syntax to include it in our simulation program (trace sink).

We are interested in including a trace source of the `MacTxBackoff` event. To get its corresponding `TracedCallback`, we can execute the following command:

```
grep 'MacTxBackoff' * -n3
```

From the results (shown in *Figure 2.19*) of this command, note down the `MakeTraceSourceAccessor m_macTxBackoffTrace` details:

```
csma-net-device.cc:127:    .AddTraceSource ("MacTxBackoff",
csma-net-device.cc-128-                     "Trace source indicating a packet has been "
csma-net-device.cc-129-                     "delayed by the CSMA backoff process",
csma-net-device.cc-130-                     MakeTraceSourceAccessor (&CsmaNetDevice::m_macTxBackoffTrace),
```

Figure 2.19 – The MacTxBackOff trace variable connecting sources details

Next, write a callback function to connect the `m_macTxBackoffTrace` trace source.

Let's write the necessary callback functions and connect them in our simulation.

From the `TracedCallback<Ptr<const Packet>> m_macTxBackoffTrace` syntax, we can define the corresponding callback function as follows in our simulation program:

```
static void MacTxBackOffTrace (std::string context, Ptr<const
Packet> packet)
{
  std::cout << context << " MAC TX back off =" <<*packet <<
std::endl;
}
```

Let's understand the arguments of our callback function before using them:

- The first argument is the context of the event (string context).

- The remaining arguments should match the number and type of arguments listed in `TracedCallback<arg1, arg2>`. As `m_macTxBackoffTrace` contains only `Ptr<const Packet> packet`, we included the same argument in our callback function.

Finally, we will use `Config::Connect` by passing our callback function to connect `MacTxBackoff` traces of all nodes in our simulation program:

```
Config::Connect ("/NodeList/*/DeviceList/*/MacTxBackoff",
MakeCallback (&MacTxBackOffTrace));
```

While using `Config::Connect`, the following details should match in your simulation program:

- `TraceSourceName: /NodeList/*/DeviceList/*/ MacTxBackoff`, which should be the exact trace source name defined in the CSMA module with the `AddTraceSource ()` line of code. Here, `/NodeList/*/DeviceList/*/` means we are interested in connecting `MacTxBackoff` traces of all the nodes and their interfaces.

- `Arguments`: Arguments of `MakeCallback` should be the address of your callback function (`&MacTxBackOffTrace`).

Similarly, we can find `MacTxOk` trace sources and connect them to the simulation program.

Now, let's write a complete trace program for inspecting CSMA-processed packets.

We can change our `pkt_first_anim.cc` (discussed in *Chapter 1*) file to write `pkt_first_trace_csma.cc`, to include the following trace-related lines of code:

- First, include the necessary header files, then define the following callback functions related to the `MacTxOk` and `MacTxBackoff` traces for saving trace details into `backoff.dat` and `txok.dat` using the `backoffStream` and `txokStream` streams:

```
std::ofstream backoffStream;
std::ofstream txokStream;
static void MacTxBackOffTrace (std::string context,
Ptr<const Packet> packet)
{
  backoffStream << Simulator::Now () <<context<<" "<<
*packet<< std::endl;
}
static void MacTxOkTrace (std::string context, Ptr<const
Packet> packet)
```

```
{
    txokStream << Simulator::Now () <<context<<" "<<
    *packet<< std::endl;
}
```

- We will modify the `main ()` function of `pkt_first_anim.cc` as follows:

  - No changes to CSMA LAN topology setup and UDP application configurations

  - We will include the following lines of code to open the `backoff.dat` and `txok.dat` files, connecting the `MacTxOk` and `MacTxBackoff` traces:

```
int
main (int argc, char *argv[])
{
    //Initial lines of code till UDP applications
configuration is same as pkt_first_anim.cc
    backoffStream.open ("backoff.dat", std::ios::out);
    txokStream.open ("txok.dat", std::ios::out);
```

- We are connecting trace sources of all nodes in our simulation using `Config::Connect`. In the callback function, using the `context` argument, it is possible to identify traces of the respective nodes:

```
    Config::Connect ("/NodeList/*/DeviceList/*/MacTx",
MakeCallback (&MacTxOkTrace));
    Config::Connect ("/NodeList/*/DeviceList/*/
MacTxBackoff", MakeCallback (&MacTxBackOffTrace));

    AnimationInterface anim ("first_trace_demo.xml");
    anim.EnablePacketMetadata ();

    Simulator::Stop (Seconds (11.0));
    Simulator::Run ();
    Simulator::Destroy ();
}
```

Now, run the simulation and observe the results using the following commands (from the `ns-allinone-3.36/ns-3.36` folder):

```
./ns3 run scratch/pkt_first_trace_csma
```

It generates two trace stream files:

- `txok.dat`
- `backoff.dat`

Let's first open `txok.dat` to check whether all packets are successfully transmitted by Node-0 or Node-3. To simplify the inspection of the packets, we can use the following `grep` commands:

```
grep 'NodeList/0' txok.dat
```

All of Node-0's UDP echo client-transmitted packets are displayed in *Figure 2.20*:

```
+2.006e+09ns/NodeList/0/DeviceList/0/MacTx ns3::EthernetHeader ( length/type=0x806, source=00:00:00:00:00:01, destination=ff:ff:ff:ff:ff:ff) ns3::ArpHea
der (request source mac: 00-06-00:00:00:00:00:01 source ipv4: 192.168.1.1 dest ipv4: 192.168.1.4) Payload (size=18) ns3::EthernetTrailer (fcs=0)
+2.00602e+09ns/NodeList/0/DeviceList/0/MacTx ns3::EthernetHeader ( length/type=0x800, source=00:00:00:00:00:01, destination=00:00:00:00:00:04) ns3::Ipv4
Header (tos 0x0 DSCP Default ECN Not-ECT ttl 64 id 0 protocol 17 offset (bytes) 0 flags [none] length: 540 192.168.1.1 > 192.168.1.4) ns3::UdpHeader (le
ngth: 520 49153 > 9) Payload (size=512) ns3::EthernetTrailer (fcs=0)
+2.00903e+09ns/NodeList/0/DeviceList/0/MacTx ns3::EthernetHeader ( length/type=0x806, source=00:00:00:00:00:01, destination=00:00:00:00:00:04) ns3::ArpH
eader (reply source mac: 00-06-00:00:00:00:00:01 source ipv4: 192.168.1.1 dest mac: 00-06-00:00:00:00:00:04 dest ipv4: 192.168.1.4) Payload (size=18) ns
3::EthernetTrailer (fcs=0)
+3e+09ns/NodeList/0/DeviceList/0/MacTx ns3::EthernetHeader ( length/type=0x800, source=00:00:00:00:00:01, destination=00:00:00:00:00:04) ns3::Ipv4Header
 (tos 0x0 DSCP Default ECN Not-ECT ttl 64 id 1 protocol 17 offset (bytes) 0 flags [none] length: 540 192.168.1.1 > 192.168.1.4) ns3::UdpHeader (length:
520 49153 > 9) Payload (size=512) ns3::EthernetTrailer (fcs=0)
+4e+09ns/NodeList/0/DeviceList/0/MacTx ns3::EthernetHeader ( length/type=0x800, source=00:00:00:00:00:01, destination=00:00:00:00:00:04) ns3::Ipv4Header
 (tos 0x0 DSCP Default ECN Not-ECT ttl 64 id 2 protocol 17 offset (bytes) 0 flags [none] length: 540 192.168.1.1 > 192.168.1.4) ns3::UdpHeader (length:
520 49153 > 9) Payload (size=512) ns3::EthernetTrailer (fcs=0)
```

Figure 2.20 – Node-0's generated UDP echo packet details

```
grep 'NodeList/3' txok.dat
```

All of Node-3's UDP echo server-transmitted packets are listed in *Figure 2.21*:

```
+2.00601e+09ns/NodeList/3/DeviceList/0/MacTx ns3::EthernetHeader ( length/type=0x806, source=00:00:00:00:00:04, destination=00:00:00:00:00:01) ns3::ArpH
eader (reply source mac: 00-06-00:00:00:00:00:04 source ipv4: 192.168.1.4 dest mac: 00-06-00:00:00:00:00:01 dest ipv4: 192.168.1.1) Payload (size=18) ns
3::EthernetTrailer (fcs=0)
+2.00903e+09ns/NodeList/3/DeviceList/0/MacTx ns3::EthernetHeader ( length/type=0x806, source=00:00:00:00:00:04, destination=ff:ff:ff:ff:ff:ff) ns3::ArpH
eader (request source mac: 00-06-00:00:00:00:00:04 source ipv4: 192.168.1.4 dest ipv4: 192.168.1.1) Payload (size=18) ns3::EthernetTrailer (fcs=0)
+2.00904e+09ns/NodeList/3/DeviceList/0/MacTx ns3::EthernetHeader ( length/type=0x800, source=00:00:00:00:00:04, destination=00:00:00:00:00:01) ns3::Ipv4
Header (tos 0x0 DSCP Default ECN Not-ECT ttl 64 id 0 protocol 17 offset (bytes) 0 flags [none] length: 540 192.168.1.4 > 192.168.1.1) ns3::UdpHeader (le
ngth: 520 9 > 49153) Payload (size=512) ns3::EthernetTrailer (fcs=0)
+3.00001e+09ns/NodeList/3/DeviceList/0/MacTx ns3::EthernetHeader ( length/type=0x800, source=00:00:00:00:00:04, destination=00:00:00:00:00:01) ns3::Ipv4
Header (tos 0x0 DSCP Default ECN Not-ECT ttl 64 id 1 protocol 17 offset (bytes) 0 flags [none] length: 540 192.168.1.4 > 192.168.1.1) ns3::UdpHeader (le
ngth: 520 9 > 49153) Payload (size=512) ns3::EthernetTrailer (fcs=0)
+4.00001e+09ns/NodeList/3/DeviceList/0/MacTx ns3::EthernetHeader ( length/type=0x800, source=00:00:00:00:00:04, destination=00:00:00:00:00:01) ns3::Ipv4
Header (tos 0x0 DSCP Default ECN Not-ECT ttl 64 id 2 protocol 17 offset (bytes) 0 flags [none] length: 540 192.168.1.4 > 192.168.1.1) ns3::UdpHeader (le
ngth: 520 9 > 49153) Payload (size=512) ns3::EthernetTrailer (fcs=0)
```

Figure 2.21 – Node-3's generated UDP echo packet details

From the screenshots, the following details are observed: simulation time, node identifier, and NIC identifier, followed by the packet contents. These details are written into `txok.dat` by our trace sink [callback function]. Next, check Node-1's and Node-2's generated list of UDP packets using the following commands:

```
grep 'NodeList/1' txok.dat
grep 'NodeList/2' txok.dat
```

As Node-1 and Node-2 do not transmit any packets, nothing will be displayed from the `txok.dat` file.

Similarly, we can observe a list of the backoff packets at various nodes during simulation execution from the `backoff.dat` file by executing the following commands:

```
grep 'NodeList/0' backoff.dat
```

All backoff packets at Node-0 are displayed in *Figure 2.22*:

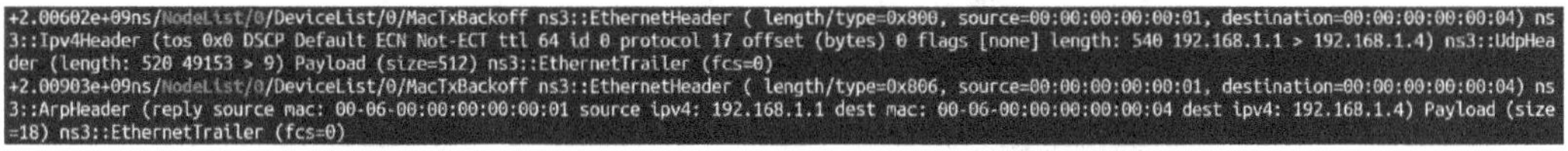

Figure 2.22 – Node-0's backoff packet details

```
grep 'NodeList/3' backoff.dat
```

All backoff packets at Node-3 are displayed in *Figure 2.23*:

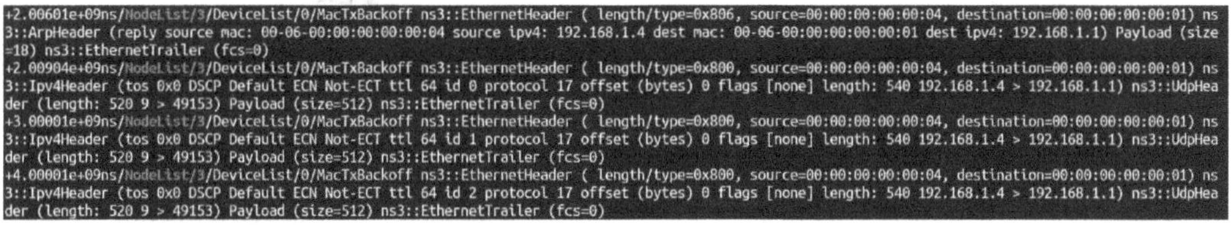

Figure 2.23 – Node-3's backoff packet details

Next, check Node-1's and Node-2's backoff packet details using the following commands:

```
grep 'NodeList/1' backoff.dat
grep 'NodeList/2' backoff.dat
```

As Node-1 and Node-2 do not attempt to transmit any packets, there are also no backoff packets.

## Understanding TCP protocol implementation in ns-3

Trace sources are highly useful to inspect the workings of protocols and collect important parameter value changes during simulation execution. In this section, we discuss an interesting hands-on task related to inspecting the working of a TCP protocol state machine and its important parameters, such as the congestion window. In the previous hands-on activity, we discussed connecting trace sources at the node level. In this hands-on activity, we'll see how to inspect the socket and its protocol working. We will also see how to inspect TCP protocol behavior and its parameters.

Let's start with identifying trace sources related to the TCP protocol. In ns-3, TCP protocol implementation is part of the `internet` module. Hence, we can find its trace source in `src/internet/model`:

```
grep 'AddTraceSource' *
```

It displays all supported trace sources of the `internet` module (shown in *Figure 2.24*). We can observe the tcp sources in `tcp-socket-base.cc`:

Figure 2.24 – The various trace sources available in the internet module

From the results shown in the screenshot, many trace sources related to the TCP protocol can be observed. In this activity, we'll also inspect TCP protocol state machine changes and congestion window changes in a simulation execution.

In order to find `TracedCallbacks` related to TCP protocol `State`, open `tcp-socket-base.cc` and check the corresponding `AddTraceSources`:

```
.AddTraceSource ("State","TCP state",MakeTraceSourceAccessor
(&TcpSocketBase::m_state),"ns3::TcpStatesTracedValueCallback")
```

Next, find the callback function related to `TcpStatesTracedValueCallback` in `tcp-socket.h`:

```
typedef void (* TcpStatesTracedValueCallback)(const
TcpSocket::TcpStates_t oldValue, const TcpSocket::TcpStates_t
newValue);
static void StateChangeTrace (std::string context, const
TcpSocket::TcpStates_t oldValue, const TcpSocket::TcpStates_t
newValue)
{
  std::cout << context << "TCP state " << oldValue << "New
value"<< newValue<< std::endl;
}
```

Now, in your simulation program, connect `StateChangeTrace` using `Config::Connect`:

```
Config::Connect ("/NodeList/*/$ns3::TcpL4Protocol/SocketList/*/
State", MakeCallback (&StateChangeTrace));
```

Similarly, we can find the `CongestionWindow` trace source and connect it in our simulation program by checking `AddTraceSource ("CongestionWindow")`, defining a callback function related to it, and connecting it in our simulation program.

Now, let's write `pkt_first_trace_tcp.cc` by changing `pkt_first_trace_csma.cc` to display TCP traces in our simulation.

After the ns-3 packages import statements, start with defining callback functions related to TCP state change and congestion window change traces:

```
std::ofstream cwndtrace;
static void
StateChangeTrace (std::string context, const
TcpSocket::TcpStates_t oldValue, const TcpSocket::TcpStates_t
newValue)
{
    std::cout << std::fixed << std::setprecision (6) <<
Simulator::Now ().GetSeconds () << std::setw (12) << context<<
oldValue << " "<<newValue << std::endl;
}
static void
TcpCwndTrace (uint32_t oldval, uint32_t newval)
{
  cwndtrace << std::fixed << std::setprecision (6) <<
Simulator::Now ().GetSeconds () << std::setw (12) << newval <<
std::endl;
}
```

In order to connect TCP trace sources in your simulation, it is necessary to wait till the corresponding sockets are available. To do this, we schedule the connecting socket traces after a certain duration, `Simulator::Schedule (MicroSeconds (1001), &ConnectSocketTraces)`:

```
Void ConnectSocketTraces (void)
{
  Config::ConnectWithoutContext ("/
NodeList/*/$ns3::TcpL4Protocol/SocketList/0/CongestionWindow",
MakeCallback (&TcpCwndTrace));
```

```
    Config::Connect ("/NodeList/*/$ns3::TcpL4Protocol/
SocketList/*/State", MakeCallback (StateChangeTrace));
}
```

In main (), there is no change in the CSMA LAN and configuration code. But, instead of a UDP application, we will install TCP bulk sender and packet sink applications:

```
int
main (int argc, char *argv[])
{
    // Initial lines of code same as pkt_first_csma_trace.cc
    uint64_t port=5000;
```

First, we will install the TCP BulkSender on Node-0 to send 1,000,000 bytes to Node-3 and configure its start and stop times accordingly:

```
    BulkSendHelper sourceHelper ("ns3::TcpSocketFactory",
                                 InetSocketAddress (interfaces.
GetAddress (3), port));
    sourceHelper.SetAttribute ("MaxBytes", UintegerValue
(1000000));
    ApplicationContainer sourceApp = sourceHelper.Install (nodes.
Get (0));
    sourceApp.Start (Seconds (0.0));
    sourceApp.Stop (Seconds (10.0));
```

Next, we will install the TCP PacketSink application on Node-3 to receive data from Node-0 and configure its start and stop times accordingly:

```
    PacketSinkHelper sinkHelper ("ns3::TcpSocketFactory",
                                 InetSocketAddress (Ipv4Address::GetAny
(), port));
    ApplicationContainer sinkApp = sinkHelper.Install (nodes.Get
(3));
    sinkApp.Start (Seconds (0.0));
    sinkApp.Stop (Seconds (10.0));
```

After the application's installation, open the `cwnd.dat` file to capture congestion window traces of Node-0, and schedule the trace events:

```
cwndStream1.open ("cwnd.dat", std::ios::out);
Simulator::Schedule (MicroSeconds (1001),
&ConnectSocketTraces);
Simulator::Stop (Seconds(11.0));
Simulator::Run ();
Simulator::Destroy ();
}
```

Finally, execute our simulation to inspect TCP state and congestion window changes during simulation execution by executing the following command:

```
./ns3 run scratch/pkt_first_trace_tcp
```

It generates a `cwnd.dat` file and displays the TCP state changes in Node-0 and Node-3, as shown in *Figure 2.25*:

```
0.009047/NodeList/0/$ns3::TcpL4Protocol/SocketList/0/State2 4
0.051371/NodeList/0/$ns3::TcpL4Protocol/SocketList/0/State4 7
0.053850/NodeList/0/$ns3::TcpL4Protocol/SocketList/0/State7 8
0.053869/NodeList/0/$ns3::TcpL4Protocol/SocketList/0/State8 10
```

Figure 2.25 – Node-0 TCP state transitions

In *Figure 2.25*, we can observe Node-0 TCP state changes (2 to 4) SYN_SENT to ESTABLISHED, (4 to 7) ESTABLISHED to FIN_WAIT1, (7 to 8) FIN_WAIT1 to FIN_WAIT2, and (8 to 10) FIN_WAIT2 to TIME_WAIT. Next, observe TCP state transitions at Node-3 in *Figure 2.26*:

```
10.000000/NodeList/3/$ns3::TcpL4Protocol/SocketList/0/State1 0
```

Figure 2.26 – The Node-3 TCP state transitions

Node-3 is a server, hence its TCP is listening at the beginning. At the end of the simulation, its TCP state is changed to a CLOSED state: [1 to 0] LISTEN to CLOSED.

Now, let's open the `cwnd.dat` file to see congestion window changes (in *Figure 2.27*) at Node-0:

```
0.009047        5360
0.009075        5896
0.009120        6432
0.009161        6968
0.009193        7504
0.009235        8040
0.009298        8576
0.009363        9112
0.009406        9648
0.009454       10184
0.009490       10720
0.009538       11256
```

Figure 2.27 – Node-0 TCP congestion window details

We can use the `cwnd.dat` file to plot congestion window changes of Node-0.

Well done. We have learned about an important feature of ns-3, tracing, to inspect the working of a network protocol and gather necessary parameter values. Now, we are going to revisit Flow Monitor in detail to evaluate the performance of simulations.

## Performance evaluation and validation of simulations using Flow Monitor

The primary objective after running a simulation is validating the results. That means we need to collect important performance metrics related to applications running in our simulation. In order to do this systematically, ns-3 offers the **Flow Monitor** module. We can install a `FlowMonitor` statistics collector application on all nodes in a simulation program. It helps the Flow Monitor module to categorize and analyze flows based on five tuples: **<source IP, source port, destination IP, destination port, protocol>**. Using the five tuples, Flow Monitor classifies every flow, and with respect to the flow, it computes throughput, delay, and packet loss metrics. For example, in a simulation, UDP echo server and client applications are installed for evaluation. Then, after the simulation execution, Flow Monitor helps you to compute two flows' results: the UDP echo server -> UDP echo client and UDP echo client -> UDP echo server performance metrics. We will discuss the following important details related to the ns-3 Flow Monitor:

- How to use Flow Monitor in ns-3 simulations
- Understanding Flow Monitor results

## How to use Flow Monitor in ns-3 simulations

In order to use the Flow Monitor features in your simulation, the following steps should be followed in the same order:

1. First, include `flow-monitor-module.h` in your list of packages. It allows you to use the `FlowMonitor` class and its features:

   ```
   #include "ns3/flow-monitor-module.h"
   ```

2. Next, after installing and configuring all necessary applications on simulation nodes, install `FlowMonitor` on all nodes using the following lines of code to collect flow-level statistics. Hence, at the end of the simulation, we can view and analyze each traffic flow's throughput, delay, jitter, and packet loss:

   ```
   Ptr<FlowMonitor> flowmon;
   FlowMonitorHelper flowmonHelper;
   flowmon = flowmonHelper.InstallAll ();
   ```

3. Next, configure the stop time of your simulation using `Simulator::Stop()`. It is necessary to stop the simulation when you use Flow Monitor; otherwise, your simulation never ends:

   ```
   Simulator::Stop (Seconds(11.0));
   Simulator::Run ();
   ```

4. Then, convert the Flow Monitor results into an XML file to parse and view the flow-level performance metrics at the end of the simulation:

   ```
   flowmon->SerializeToXmlFile ("first_flowmetrics.xml",
   true, true);
   ```

5. After making the preceding changes in your simulation program, save and execute the simulation. It generates the Flow Monitor results XML file.

6. Finally, copy the `ns-allinone-3.36/ns-3.36/src/flow-monitor/examples/flowmon-parse-results.py` file to the `scratch` folder.

7. Execute the `python3 scratch/flowmon-parse-results.py first_flowmetrics.xml` command and check the Flow Monitor results.

8.  It is also possible to view the Flow Monitor results (refer to *Figure 2.28*) in the **NetAnim** visualizer under the **Stats** tab. From the **Stats** tab, select **Flow-monitor**, then click the **FlowMon file** button and open `first_flowmetrics.xml` generated by your simulation program:

Figure 2.28 – Flow Monitor results visualization in NetAnim

Now, we can view flow-level stats of all nodes or individual nodes by clicking on the node numbers displayed in the side panel. We selected the **ALL** option to show all flows' statistics.

## Understanding Flow Monitor results

After parsing the Flow Monitor-generated XML file, we can check flow-level statistics. While computing flow-level statistics, Flow Monitor assigns a unique ID to each flow based on the five tuples and computes the flow-level performance statistics. Hence, after parsing the XML file, it displays each flow's results in terms of FlowID, Tx bitrate, Rx bitrate, Mean Delay, and Packet Loss Ratio.

For example, in our `pkt_first.cc` simulation, by configuring UDP client traffic characteristics (MaxPackets: 100, Interval: 0.1, and PacketSize: 1,024 bytes), the following UDP results (shown in *Figure 2.29*) are obtained:

```
FlowID: 1 (UDP 192.168.1.1/49153 --> 192.168.1.2/9)
        TX bitrate: 85.23 kbit/s
        RX bitrate: 85.23 kbit/s
        Mean Delay: 1.01 ms
        Packet Loss Ratio: 0.00 %
FlowID: 2 (UDP 192.168.1.2/9 --> 192.168.1.1/49153)
        TX bitrate: 85.23 kbit/s
        RX bitrate: 85.23 kbit/s
        Mean Delay: 1.01 ms
        Packet Loss Ratio: 0.00 %
```

Figure 2.29 – The Flow Monitor results

Let's understand these results thoroughly.

As in the simulation, there are two UDP flows. Results for each of the flows are displayed with its flow ID and its five tuple details. For example, we can observe flow 1's details:

```
FlowID-1 (Protocol: UDP SourceIP: 192.168.1.1 SourcePort:49153,
DestinationIP: 192.168.1.2, DestinationPort: 9)
```

Similarly, flow 2's details can be interpreted.

Now, we'll look at flow 1's performance metrics:

- Tx bitrate: Number of bits successfully transmitted per second from 192.168.1.1/49153 to 192.168.1.2/9. In our result, it is 85.23 kbps and the result matches the UDP client sending rate (10*1024*8/sec).

- Rx bitrate: Number of bits successfully received per second from 192.168.1.1/49153 to 192.168.1.2/9. As all transmitted bits are successfully received, we observe an Rx rate of 85.23 kbps. If the UDP client traffic sending rate is greater than the link speed, you will observe a lower Rx bitrate due to packet losses.

- Mean Delay: The delay of a packet is the sum of its transmission delay and propagation delay. Flow Monitor computes the average delay of all packets received. In our simulation, it is 1.01 ms.

- Packet Loss Ratio: This is computed as a ratio of the number of packets lost to the number of packets transmitted. Since in our simulation there are no packets lost, it is 0%.

Flow 2 represents the UDP echo server. Since the UDP client successfully transmitted all its packets to the UDP echo server, the UDP echo server is able to reply to all these packets successfully. Hence, flow 2's results are the same as flow 1's.

Let's change the UDP client sending rate to 8 Mbps (set Interval to 0.001 and PacketSize to 1024 bytes), MaxPackets to 100000, and configure the P2P link speed (Data Rate) to 5 Mbps in pkt_first.cc. After running the simulation, parse the Flow Monitor XML file and check the results in *Figure 2.30*:

```
FlowID: 1 (UDP 192.168.1.1/49153 --> 192.168.1.2/9)
        TX bitrate: 8417.05 kbit/s
        RX bitrate: 4991.36 kbit/s
        Mean Delay: 1235.74 ms
        Packet Loss Ratio: 26.68 %
FlowID: 2 (UDP 192.168.1.2/9 --> 192.168.1.1/49153)
        TX bitrate: 4991.56 kbit/s
        RX bitrate: 4991.55 kbit/s
        Mean Delay: 2.69 ms
        Packet Loss Ratio: 0.00 %
```

Figure 2.30 – Flow Monitor results after changing the traffic configuration

Let's understand the Flow Monitor results shown in *Figure 2.30*:

- For flow 1, we can observe Tx bitrate is 8,417 Kbps (8 Mbps), but Rx bitrate is 4,991 Kbps (around 5 Mbps). The lower Rx bitrate is due to the UDP client application sending packets at a higher speed than the link speed (a Data Rate value of 5 Mbps) and it is leading to many packets being lost at the sender side itself. Hence, we can observe that Packet Loss Ratio is 26.68%. It is also resulting in an increased Mean Delay.

- Flow 2 corresponds to the UDP echo server. Hence, its sending rate is now only 4,991 kbps. As the sending rate is lower than or equal to the P2P Data Rate, there is no packet loss (0%), and `Tx bitrate` and `Rx bitrate` are the same (4,991 Kbps). Besides, we can also observe a lower `Mean Delay` for flow 2 compared to flow 1.

We now understand the UDP results. Similarly, we can also view the TCP flow results. The only difference is that Flow Monitor gives results for the TCP flows of ACKs also. Hence, while projecting results, we should check TCP data flows results carefully by separating TCP ACK flow results.

Now we understand the Flow Monitor results thoroughly. Next, we'll learn how to use `gnuplot` to plot simulation results.

# Using gnuplot to project simulation results

One of the interesting activities in analyzing simulation results is projecting results visually using a variety of plots. In this section, we discuss how to project simulation results using the `gnuplot` tool. In ns-3, simulation results are usually obtained using Flow Monitor. But the format of Flow Monitor results may not be suitable for plotting. Flow Monitor results need to be parsed and converted into suitable input files for `gnuplot`. In order to carry out these activities, we explain simple parsing procedures using Unix commands such as `grep`, `awk`, `cut`, and `paste`. In this section, related to projecting simulation results, we discuss the following:

- The `gnuplot` basics
- How to parse simulation results for plotting
- How to plot line charts and bar charts

## The gnuplot basics

Users who are new to `gnuplot` will first need to install it using the `apt-get install gnuplot` command and go through the following `gnuplot` commands before using it:

| Command | Description |
| --- | --- |
| `set terminal` | To set the output file type and font details |
| `set output` | To set the output file name, for example, `plot1.jpg` |
| `set title` | To set the plot title |
| `set xlabel` | To set the x-axis title |
| `set xtics` | To define the x-axis step size |
| `set ylabel` | To set the y-axis title |
| `set ytics` | To define the y-axis step size |

| Command | Description |
|---|---|
| set xrange | To set the x-axis tics range |
| set yrange | To set the y-axis tics range |
| set style | To set the style of the plot, such as lines, line points, boxes, and fill patterns |
| plot | To plot any chart using the input data file |

## How to parse simulation results for plotting

Run the following commands from your ns-3 installed folder (ns-allinone-3.36/ns-3.36):

1. First, extract the flow IDs into the flows file:

```
python3 scratch/flowmon-parse-results.py firstanim_
flowmetrics.xml|grep 'FlowID' |cut -f2 -d ' ' > flows
```

2. Next, create the flowsvsrxthrput file, which contains the flow ID and corresponding Tx bitrate:

```
python3 scratch/flowmon-parse-results.py firstanim_
flowmetrics.xml|grep 'TX bitrate' | cut -f3 -d ' '
>txthrput
paste flows txthrput > flowsvstxthrput
cat flowsvstxthrput
1    85.23
2    85.29
```

3. Next, create the flowsvsrxthrput file, which contains the flow ID and corresponding Rx bitrate:

```
python3 scratch/flowmon-parse-results.py firstanim_
flowmetrics.xml|grep 'RX bitrate' | cut -f3 -d ' '
>rxthrput
paste flows rxthrput > flowsvsrxthrput
cat flowsvsrxthrput
1    85.23
2    85.29
```

4.  Next, create the `flowsvsplr` file, which contains the flow ID and corresponding `Packet Loss Ratio`:

```
python3 scratch/flowmon-parse-results.py firstanim_
flowmetrics.xml|grep 'Packet Loss Ratio' |cut -f4 -d ' '
>plr
paste flows plr > flowsvsplr
cat flowsvsplr
1  0.00
2  0.00
```

5.  Finally, create the `flowsvsdelay` file, which contains the flow ID and corresponding `Mean Delay`:

```
python3 scratch/flowmon-parse-results.py firstanim_
flowmetrics.xml|grep 'Mean Delay' |cut -f3 -d ' ' > delay
paste flows delay > flowvsdelay
cat flowsvsdelay
1  1.80
2  1.76
```

Oh, that's great. We have successfully parsed the Flow Monitor stats file, `firstanim_flowmetrics.xml`, to collect flow-level performance metrics into corresponding input data files, such as `flowsvstxthrput`, `flowsvsrxthrput`, `flowsvsplr`, and `flowsvsdelay`. Now, we can use these files for plotting using gnuplot.

## How to plot line charts and bar charts

Let's see how to plot flow ID against `Tx Throughput` using the `linespoints` chart. Write the following commands in the `txthrplot` file and save it:

```
set terminal postscript eps color enh "Times-BoldItalic"
set output "flowsvstxthrput.eps"
set title "Flows vs Tx Throughput"
set xlabel "Flows "
set ylabel "Tx Throughput (kbps)"
set xrange [0:3]
set xtics 1
set yrange [0:100]
set style line 1 linecolor 3 linetype 1 linewidth 2 pointtype 5
pointsize 1.5
```

```
plot 'flowsvstxthrput' using 1:2 title 'FlowID vs Tx
Throughput' with linespoints linestyle 1
```

Now, execute the following command to generate the flowsvstxthrput.eps file (refer to *Figure 2.31*):

**gnuplot txthrplot**

It generates the plot in flowsvstxthrput.eps. We can open it using a PDF viewer:

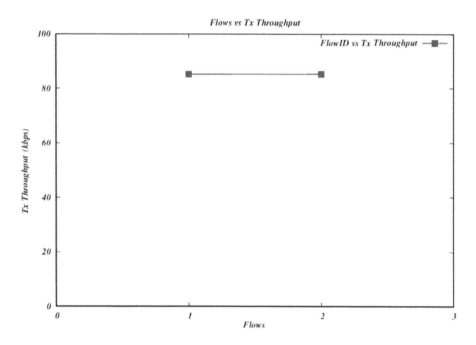

Figure 2.31 – Transmitted throughput (Flows versus Tx Throughput (kbps))

Now, let's see how to plot flow ID against Rx Throughput using a bar or boxes chart. Write the following commands in the rxthrplot file and save it:

```
set terminal postscript eps color enh "Times-BoldItalic"
set output "flowvsrxthr.eps"
set title "Flow vs Rx Throughput"
set xlabel "Flows "
set ylabel "Rx Throughput (Kbps)"
set xrange [0:3]
set xtics 1
```

```
set yrange [0:100]
set boxwidth 0.1
set style fill pattern 2
plot "flowsvsrxthrput" using 1:2 title 'FlowID vs Rx
Throughput' with boxes lc 2
```

Now, run the following command to generate a bar chart. It generated the `flowvsrxthr.eps` output file (refer to *Figure 2.32*):

**gnuplot rxthrplot**

It generates the plot in `flowsvsrxthr.eps`. We can open it using a PDF viewer:

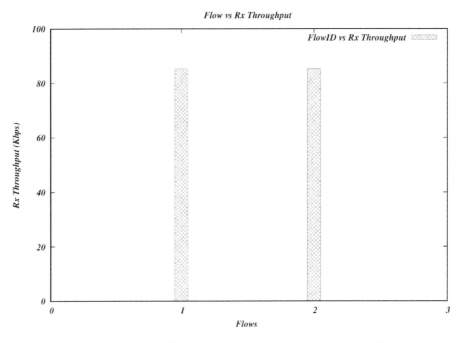

Figure 2.32 – Received throughput (Flows versus Rx Throughput (Kbps))

As both flows' throughputs are the same, we can observe that flow 1's and flow 2's bars have the same height.

# Summary

Congratulations on successfully learning about ns-3's important features! This will help you to handle complex simulation information, errors, and crashes using logging and debugging.

After learning about the logging and debugging features, you will feel comfortable with handling a variety of simulation errors or crashes quickly. In particular, learning how to use the ns-3 tracing feature is highly useful to understand various protocols and inspect important parameters at various levels. ns-3 tracing simplifies how to use various modules nodes, network devices, protocols, and algorithms in user simulations without making any changes to existing modules.

As you've learned how to use Flow Monitor in your simulations, this will save you a lot of time in collecting network performance metrics such as throughput, delay, and packet loss details for validating simulation results. As part of this chapter, besides learning how to validate simulation results, you also learned how to project your simulation results systematically using a variety of plots using `gnuplot`.

In the next chapter, we are going to learn about the key building blocks of ns-3, such as a variety of nodes, net devices, channels, and applications to simulate wired and wireless networks.

# 3

# ns-3 Key Building Blocks for Simulations

In previous chapters, we practiced using the basic features of ns-3 to quickly write simulations. We also understood the ns-3 simulation program structure, and how to set up simple topologies using important ns-3 classes related to nodes, channels, and protocols. This chapter discusses in detail key building blocks for setting up wired or wireless network topologies and configuring them using interesting network applications. This chapter starts with discussing how to create hosts such as computers, servers, and **Wi-Fi Stations** (**Wi-Fi STAs**), and a variety of network-connecting devices, such as switches, routers, and Wi-Fi **Access Points** (**APs**). In ns-3, any networking host or interconnecting device can be created using ns-3 nodes. As part of the first section, you will learn how to create and configure a variety of networking hosts or servers and interconnect devices using step-by-step procedures.

The next section of this chapter discusses how to connect networking hosts and interconnecting devices using wired and/or wireless channels to set up a variety of network topologies. Specifically, you will learn how to configure various channels, such as **Point-to-Point** (**P2P**), CSMA/CD, LTE, and Wi-Fi, and install necessary protocols for setting up wired and/or wireless topologies.

In order to test and evaluate the performance metrics of a simulation setup, it is necessary to install suitable network applications and generate traffic flows. In past chapters, you were already introduced to a few basic network applications, such as UDP and TCP applications. In this chapter, a greater variety of TCP, UDP, and HTTP network applications and their configuration details are discussed to use them in our simulations.

The final sections of this chapter discuss two important hands-on activities related to setting up basic wired **Local Area Networks** (**LANs**) and Wi-Fi LANs using step-by-step procedures. Mainly, this chapter discusses the following topics:

- Setting up a variety of network nodes using ns-3 nodes and containers
- Connecting nodes using a variety of channels

- Quickly installing and configuring InternetStack and applications
- The step-by-step procedure for building a simple LAN
- The step-by-step procedure for building a simple wireless LAN

# Setting up a variety of network nodes using ns-3 nodes and containers

ns-3 supports a wide range of wired and wireless network simulations. In order to simplify setting up wired or wireless topologies, ns-3 offers a simplified procedure to set up any topology using the simple NodeContainer, NetDeviceContainer, and a variety of communication channels. ns-3 allows users to create a large number of nodes on servers with abundant computation and memory resources. In this section, we will start with learning how to set up the following networking nodes:

- A variety of wired networking hosts such as computers and servers
- A variety of wired interconnecting devices such as switches and routers
- A variety of wireless stations such as laptops and mobile phones
- A variety of wireless base stations such as Wi-Fi **APs** and cellular base stations

## ns-3 wired networking hosts

In ns-3 simulation, computers, hosts, or servers can be created using the ns-3 Node (*Figure 3.1*) or NodeContainer class. In ns-3, creating and configuring a computer or server involves the following steps:

1. Create the necessary number of nodes using the ns-3 NodeContainer class.
2. Then, install the TCP/IP stack of nodes using the ns-3 InternetStackHelper class.
3. Finally, install necessary ns-3 network applications over ns-3 nodes:

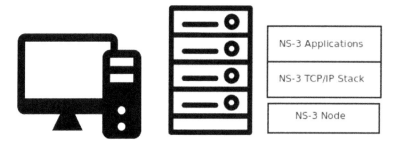

Figure 3.1 – An ns-3 computer or server

Let's create 10 hosts and 10 servers in ns-3. As we need two different sets of nodes, it is done using two `NodeContainer`, as shown in the following code snippet:

```
NodeContainer hosts;
hosts.Create(10);
NodeContainer server;
server.Create(10);
```

In order to connect these nodes in a network and to install any network applications, it is necessary to install a TCP/IP protocol stack as follows:

```
InternetStackHelper internet;
internet.Install (hosts);
internet.Install (servers); (or)
NodeContainer hosts_servers = NodeContainer (hosts,servers);
hosts_servers.install(hosts_servers)
```

Next, we will see how to set up wired interconnecting network devices to connect various nodes.

## Interconnecting networks

In general, ns-3 supports switches or routers to interconnect networking hosts. We'll start with learning how to create a simple switch (*Figure 3.2*) using the ns-3 `BridgeHelper` class, as follows:

1.  First, create a node using the ns-3 `NodeContainer` class for configuring a switch.

2.  Connect nodes using communication channels, such as CSMA, and collect switch/bridge-connecting network devices, then attach the network devices to the bridge. For example, when a node is connected to a switch using a CSMA channel, it returns two network devices (or NICs). One network device is attached to the node, and another is attached to the switch.

3.  Then, an ns-3 packet switch or bridge can be created using an ns-3 `BridgeHelper` class by installing bridge protocols on an ns-3 node attached to network devices:

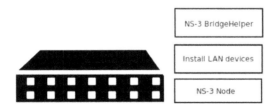

Figure 3.2 – An ns-3 switch or bridge

Let's see how to create a switch/bridge in ns-3 using the following code snippet:

```
#include "ns3/bridge-helper.h"
Ptr<Node> bridge1 = CreateObject<Node> ();
BridgeHelper bridge;
bridge.Install (bridge1, netdevices);
```

Here `bridge1` is created to simulate a switch, and its ports are connected to nodes using a communication channel such as CSMA. As the switch helps with packet switching within a network only, all nodes connected to the switch must belong to the same subnet IP. Hence, ns-3 bridges are helpful to create simple **LANs** only. In the case of interconnecting LANs in a larger network, it is necessary to deploy routers in the network. Next, we will learn how to set up a router to connect different networks.

### Simple router setup and configuration in ns-3

Let's see how to create and set up a router (*Figure 3.3*) in ns-3 using the following procedure:

1.  First, create a node using the ns-3 `NodeContainer` class.
2.  Then, install the internet stack using the `InternetStackHelper` class.
3.  Finally, configure routing algorithms on router nodes.

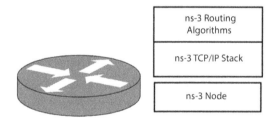

Figure 3.3 – ns-3 router setup

In ns-3, a router can be created using the following line of code:

```
NodeContainer router;
router.Create(1);
InternetStackHelper internet;
internet.Install (router);
Ipv4GlobalRoutingHelper::PopulateRoutingTables ();
```

Till now, we have seen how to create and set up wired networking devices. Next, we'll learn how to create and set up wireless networking devices.

## Wireless networking devices setup

In ns-3, a Wi-Fi-enabled laptop or mobile phone is referred to as a Wi-Fi STA. First, let's see how to create Wi-Fi STAs (*Figure 3.4*) using ns-3 nodes. In ns-3, any Wi-Fi STA can be created and configured as follows:

1.  Create the necessary number of Wi-Fi STAs using the ns-3 `NodeContainer` class.
2.  Next, create a Wi-Fi channel and install the ns-3 Wi-Fi **Physical** (**PHY**) layer.
3.  Then, configure the Wi-Fi MAC protocol and install the ns-3 Wi-Fi **Media Access Control (MAC)** layer.
4.  Finally, install the ns-3 TCP/IP stack.

| ns-3 Application |
| ns-3 TCP/IP Stack |
| ns-3 Wi-Fi MAC |
| ns-3 Wi-Fi PHY |
| ns-3 Node |

Figure 3.4 – The ns-3 Wi-Fi STAs and APs

To create wireless network topologies, the following ns-3 Wi-Fi packages are needed in your simulation program:

```
#include "ns3/yans-wifi-helper.h"
#include "ns3/yans-wifi-channel.h"
#include "ns3/ssid.h"
#include "ns3/mobility-helper.h"
```

Let's see how to create Wi-Fi STAs using the following ns-3 code snippets.

Create the necessary Wi-Fi STAs using ns-3 `NodeContainer` in `main()`:

```
NodeContainer stas;
NetDeviceContainer staDevs;
stas.Create (10);
```

In order to enable Wi-Fi features on these stations, Wi-Fi PHY-layer and MAC-layer protocols should be installed. These steps can be done using the following ns-3 lines of code.

First, carry out basic Wi-Fi PHY-layer protocol configuration using the following code. It involves creating a Wi-Fi channel:

```
YansWifiPhyHelper wifiPhy;
YansWifiChannelHelper wifiChannel =
YansWifiChannelHelper::Default ();
wifiPhy.SetChannel (wifiChannel.Create ());
```

Then, carry out basic Wi-Fi MAC-layer protocol configuration using the following code. It involves setting up a **Service Set Identifier** (**SSID**) and connecting to a Wi-Fi AP using `WifiMacHelper`. ns-3 STAs support both active-probing and passive-scanning Wi-Fi connection establishment procedures. As an example, we configured `ActiveProbing` for STAs to connect with an AP:

```
WifiMacHelper wifiMac;
Ssid ssid = Ssid ("wifi-default");
wifiMac.SetType ("ns3::StaWifiMac",
                "ActiveProbing", BooleanValue (true),
                "Ssid", SsidValue (ssid));
```

Finally, install Wi-Fi PHY- and MAC-layer protocols on all Wi-Fi STAs using the ns-3 `WifiHelper` class:

```
WifiHelper wifi;
staDevs = wifi.Install (wifiPhy, wifiMac, stas);
```

Next, we will learn how to create and configure a Wi-Fi AP in ns-3.

### How to set up a Wi-Fi AP

Now, we can see how to set up a basic Wi-Fi AP. Like Wi-Fi STAs, it also involves creating necessary AP nodes, then configuring Wi-Fi PHY- and MAC-layer protocols, and finally, installing Wi-Fi PHY and MAC layers on AP nodes:

```
NodeContainer ap;
ap.Create (1);
```

Start with Wi-Fi PHY-layer protocol configuration using the following ns-3 code snippet. It involves creating a Wi-Fi channel using `YansWifiPhyHelper` and `YansWifiChannelHelper`:

```
YansWifiPhyHelper wifiPhy;
YansWifiChannelHelper wifiChannel =
```

```
YansWifiChannelHelper::Default ();
wifiPhy.SetChannel (wifiChannel.Create ());
```

Then, carry out basic Wi-Fi MAC-layer protocol configuration using the following ns-3 code snippet. It involves setting up an SSID. In doing so, Wi-Fi STAs in your simulation can connect to this AP using the same SSID:

```
WifiMacHelper wifiMac;
Ssid ssid = Ssid ("wifi-default");
wifiMac.SetType ("ns3::ApWifiMac",
                "Ssid", SsidValue (ssid));
```

Finally, install Wi-Fi PHY- and MAC-layer protocols on all Wi-Fi APs using the ns-3 `WifiHelper` class:

```
WifiHelper wifi;
staDevs = wifi.Install (wifiPhy, wifiMac, stas);
```

That's great! We have learned how to set up, including creating and configuring, a variety of wired and Wi-Fi networking devices in ns-3. Next, to show how simple it is to set up any network simulations in ns-3, we will introduce how to create 4G cellular networking devices using ns-3.

### Basic 4G cellular networking devices deployment using ns-3

Let's see how simple it is to create 4G or LTE base stations (eNodeBs) and mobile phones (**User Equipment (UE)**) in ns-3.

Let's create 10 LTE eNodeBs and 10 LTE UEs:

```
NodeContainer eNodeBs;
NodeContainer UEs;
eNodeBs.Create (10);
UEs.Create (10);
```

Now, install the LTE stack on these nodes and collect the respective network devices:

```
Ptr<LteHelper> lteHelper = CreateObject<LteHelper> ();
NetDeviceContainer enbDevs;
NetDeviceContainer ueDevs;
enbDevs = lteHelper->InstallEnbDevice (enbNodes);
ueDevs = lteHelper->InstallUeDevice (ueNodes);
```

Hey, congratulations! We have learned how to create and set up wired, wireless, and 4G networking devices in ns-3. Next, we will see how to connect them using communication channels.

# Connecting nodes using a variety of channels

ns-3 offers a variety of wired and wireless channels to connect networking nodes. In this section, we will discuss the following ns-3 channels and how to use them:

- ns-3 wired channels
- ns-3 wireless channels

## ns-3 wired channels

ns-3 supports the following wired channels to connect various networking nodes, such as hosts, computers, servers, switches, and routers:

- ns-3 CsmaChannel
- ns-3 PointToPointChannel

### ns-3 CsmaChannel

The ns-3 CSMA/CD model is helpful for simulating Ethernet or wired LANs. In ns-3, the CSMA/CD protocol is simulated for accessing a shared medium and detecting collisions. It is implemented by the ns-3 scheduler by giving access to nodes based on the **First Come First Served** (**FCFS**) principle. In a CSMA/CD network simulation, collisions are detected using a global shared variable among nodes and it also simulates random exponential backoff by nodes to retry their transmission during any collision detection. In order to connect ns-3 nodes in an Ethernet network, ns-3 offers a CsmaChannel to simulate transmission medium behavior. Any number of nodes can be connected using a CsmaChannel. Users can configure important parameters of the CSMA channel, such as link speed and propagation delay, using the DataRate and Delay attributes, respectively.

DataRate indicates the data transmission speed (bps) of the nodes connected using a CsmaChannel object. Delay indicates a combination of a signal propagation delay from a node through its CsmaChannel to the switch, the packet forwarding time at the switch, and the signal propagation time to the destination node's network device.

Now, let's see how to set up a packet switch and connect 24 hosts to it, as shown in the following topology (*Figure 3.5*), with a CSMA configuration link speed of 1 Gpbs and a propagation delay of 1 ms:

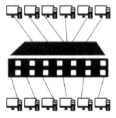

Figure 3.5 – Setting up a switch and connecting hosts to it using CSMA channels

Let's see how to set up a bridge using the CSMA module using the following procedure.

Start with importing necessary packages:

```
#include "ns3/bridge-helper.h"
```

Then, create 24 hosts using ns-3 NodeContainer using the following code snippet:

```
NodeContainer hosts;
hosts.Create(24);
```

Next, create a node for installing bridge devices and protocols to do packet switching using the following code snippet:

```
Ptr<Node> bridge = CreateObject<Node> ();
```

After creating the necessary ns-3 nodes, create a CSMA channel using ns-3's CsmaHelper and configure its DataRate and Delay parameters:

```
CsmaHelper csma;
csma.SetChannelAttribute ("DataRate", StringValue("1Gbps"));
csma.SetChannelAttribute ("Delay", StringValue("1ms"));
```

Then, connect all hosts to the bridge node using the CSMA channel and collect host network devices and bridge network devices separately to install them on the bridge using the following ns-3 code snippet:

```
NetDeviceContainer HostDevices;
NetDeviceContainer BridgeDevices;
for (int i = 0; i < 24; i++)
{
    NetDeviceContainer link = csma.Install (NodeContainer
(lan1.Get (i), bridge1));
    HostDevices.Add (link.Get (0));
    BridgeDevices.Add (link.Get (1));
    }
```

Finally, create the bridge by installing bridge devices and bridge protocols to do packet switching. Now, this bridge works similar to a switch, and all hosts are connected to it:

```
BridgeHelper bridge;
bridge.Install (bridge, lan1BridgeDevices);
```

Next, we'll see how to connect two ns-3 nodes using a dedicated channel.

## ns-3 PointToPointChannel

ns-3 offers a **P2P** model to connect exactly two ns-3 nodes using a P2P channel. An ns-3 P2P channel is a simulation of a full-duplex RS-232 or RS-422 link with null modem and no handshaking. ns-3 nodes connected using a P2P channel transmit their data using P2P protocol encapsulation. ns-3 P2P implementation assumes the P2P link was always established and authenticated. Since P2P nodes' transmission data is not framed, in their packet's `Address` and `Control` fields, error detection fields are not included. Similar to the CSMA/CD channel, users can configure the P2P channel's important parameters, such as link speed and propagation delay, using the `DataRate` and `Delay` attributes, respectively. `DataRate` indicates the data transmission speed (bps) of the nodes connected using a P2P channel. `Delay` indicates the signal propagation time from the source node's transmission medium endpoint to the destination node's network device:

Figure 3.6 – A P2P network

Let's create a P2P network for connecting two hosts (*Figure 3.6*) with a link speed of 1 Gbps and a propagation delay of 1 ms using the following code snippet:

```
NodeContainer nodes;
nodes.Create (2);
PointToPointHelper pointToPoint;
pointToPoint.SetDeviceAttribute ("DataRate", StringValue
("1Gbps"));
pointToPoint.SetChannelAttribute ("Delay", StringValue
("1ms"));
NetDeviceContainer devices;
devices = pointToPoint.Install (nodes);
```

Next, let's see how to set up a star topology using P2P links (*Figure 3.7*):

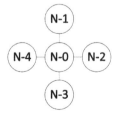

Figure 3.7 – Star topology set up using P2P links

A P2P module also supports creating multiple P2P link topologies using a P2P layout model. For example, using the following code snippet, let's create a star network for connecting five hosts, as shown in *Figure 3.7*, with a link speed of 1 Gbps and a propagation delay of 1 ms:

```
#include "ns3/point-to-point-module.h"
#include "ns3/point-to-point-layout-module.h"
 NS_LOG_INFO («Build star topology.»);
   PointToPointHelper pointToPoint;
   pointToPoint.SetDeviceAttribute ("DataRate", StringValue
("1Gbps"));
   pointToPoint.SetChannelAttribute ("Delay", StringValue
("1ms"));
```

Our topology can be created as follows. It has one central hub and four nodes as spokes:

```
    PointToPointStarHelper star (4, pointToPoint);
    NS_LOG_INFO («Install internet stack on all nodes.»);
    InternetStackHelper internet;
    star.InstallStack (internet);
```

Next, let's learn how to connect Wi-Fi nodes in ns-3 using wireless channels.

## ns-3 wireless channels

ns-3 supports various Wi-Fi standards, such as 802.11a/b/e/g/n/ac/ax specifications, and models the respective standard PHY-layer and MAC-layer features. We will discuss a variety of Wi-Fi modules in detail in upcoming chapters. In this section, you will learn how to connect Wi-Fi STAs and Wi-Fi APs using standard Wi-Fi channels. The ns-3 Wi-Fi module offers `YansWifiChannel` to connect Wi-Fi networking devices such as Wi-Fi APs and STAs. ns-3's `YansWifiChannel` simulates both signal propagation loss and propagation delay features in a Wi-Fi deployment environment.

The ns-3 Wi-Fi module offers separate MAC layers corresponding to STAs and APs. The ns-3 `StaWifiMac` class simulates active probing and passive scanning mechanisms to connect a Wi-Fi AP. On other hand, the ns-3 `ApWifiMac` class simulates important AP functionalities, such as generating periodic beacons and handling association requests from Wi-Fi STAs.

In a WLAN, STAs can connect to an AP using either active probing or passive scanning. An STA's default association mechanism is passive scanning. Besides, ns-3 Wi-Fi STAs or APs simulate the standard 802.11 **Distributed Coordination Function** (**DCF**) to get channel access before any data transmission. Before learning about ns-3 Wi-Fi modules in detail, let's learn how to set up a Wi-Fi LAN quickly using the following code snippets:

- Create an AP and configure its PHY and MAC layers. Set a particular SSID to the AP.

- Set up 10 Wi-Fi STAs by configuring their PHY and MAC layers and allowing them to connect with a particular SSID.

First, import all the necessary Wi-Fi modules:

```
#include «ns3/yans-wifi-helper.h"
#include «ns3/ssid.h"
#include «ns3/yans-wifi-channel.h"
```

Next, create the necessary ns-3 nodes to set up the Wi-Fi AP and STAs:

```
NodeContainer APs;
NetDeviceContainer apDevices;
APs.Create (1);
NodeContainer sta;
sta.Create (10);
```

After creating the necessary ns-3 nodes, set up a common Wi-Fi channel to connect the Wi-Fi AP and STAs:

```
NetDeviceContainer staDev;
NetDeviceContainer apDev;
YansWifiPhyHelper wifiPhy;
wifiPhy.SetPcapDataLinkType (WifiPhyHelper::DLT_IEEE802_11_
RADIO);
YansWifiChannelHelper wifiChannel =
YansWifiChannelHelper::Default ();
wifiPhy.SetChannel (wifiChannel.Create ());
```

Then, configure the AP's specific Wi-Fi MAC and set an SSID (e.g., Packt):

```
WifiHelper wifi;
WifiMacHelper wifiMac;
Ssid ssid = Ssid ("Packt");
wifiMac.SetType ("ns3::ApWifiMac",
                "Ssid", SsidValue (ssid));
```

Now, install the AP's Wi-Fi MAC and PHY layers onto the ns-3 AP node:

```
apDev = wifi.Install (wifiPhy, wifiMac, APs.Get (0));
```

Finally, configure the STAs' Wi-Fi MAC and PHY layers and install them on the ns-3 STA node:

```
wifiMac.SetType ("ns3::StaWifiMac",
                 "Ssid", SsidValue (ssid));
staDev = wifi.Install (wifiPhy, wifiMac, sta);
```

That's superb. We have learned how to set up wired or wireless networking devices and connect them using the respective channels. Let's quickly explore how to set up a basic 4G cellular network in ns-3.

Let's create one eNodeB and one UE using the following code snippet:

```
#include "ns3/lte-module.h"
  NodeContainer eNodeBs;
  NodeContainer UEs;
  eNodeBs.Create (1);
  UEs.Create (10);
```

Now, install the LTE stack on these nodes and collect the respective network devices:

```
Ptr<LteHelper> lteHelper = CreateObject<LteHelper> ();
  NetDeviceContainer enbDevs;
  NetDeviceContainer ueDevs;
  enbDevs = lteHelper->InstallEnbDevice (eNodeBs);
  ueDevs = lteHelper->InstallUeDevice (UEs);
```

Next, we see how to attach UEs to eNodeBs using LTE communication channels:

```
lteHelper->Attach (ueDevs, enbDevs.Get (0));
```

That's great. We have learned how to set up wired or wireless networks using the respective channels. Next, we will learn about various network applications supported by ns-3 and how to use them in our simulation.

## Quickly installing and configuring InternetStack and applications

ns-3 supports a rich set of network applications to easily use in simulations. In this section, you'll learn about the following interesting network applications that are implemented using either TCP or UDP:

- ns-3 UDP client and server applications
- ns-3 UDP echo client and server applications

- ns-3 TCP bulk send and packet sink applications

- ns-3 On-Off TCP/UDP Applications

- ns-3 3GPP HTTP client and server applications

## ns-3 UDP client and server applications

In this section, we'll learn how to simulate and test a UDP client and server application (as shown in *Figure 3.8*), where the client application sends a maximum *N* number of packets per second (e.g., 10 packets per second). However, the server application does not send any responses to the client application. To simulate this behavior in ns-3, the `UdpServerHelper` and `UdpClientHelper` classes are used:

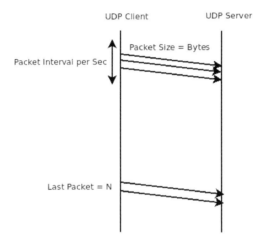

Figure 3.8 – ns-3 UDP client and server traffic flow exchange

Let's see how to simulate the UDP client and server behavior in ns-3 using the following code snippet. First, create and configure a UDP server application with a port number using the `UdpServerHelper` class:

```
uint16_t port = 4000;
UdpServerHelper server (port);
```

Next, install the UDP server application on a node with index-1. Configure its start and stop timings using the ns-3 `ApplicationContainer` class. It is always better to set the server's stop time greater than all its client's stop time values. Otherwise, the server may not receive all packets from its clients:

```
ApplicationContainer apps = server.Install (nodes.Get (1));
apps.Start (Seconds (1.0));
apps.Stop (Seconds (10.0));
```

Now, create and configure a UDP client application with the UDP server address and port number as parameters using the ns-3 UdpClientHelper class. We need to configure UDP client traffic characteristics such as MaxPackets, Interval, and PacketSize:

```
uint32_t MaxPacketSize = 1024;
Time interPacketInterval = Seconds (0.05);
uint32_t maxPacketCount = 320;
UdpClientHelper client (interfaces.GetAddress(1), port);
client.SetAttribute ("MaxPackets", UintegerValue
(maxPacketCount));
client.SetAttribute ("Interval", TimeValue
(interPacketInterval));
client.SetAttribute ("PacketSize", UintegerValue
(MaxPacketSize));
```

Finally, install the UDP client application on a node with index-0. Configure its start and stop timings using ApplicationContainer. We need to set the client's start time to greater than the server's start time, and the client's stop time should be set according to MaxPackets and the packets' interval rate. Otherwise, the client application cannot send all its packets:

```
apps = client.Install (nodes.Get (0));
apps.Start (Seconds (2.0));
apps.Stop (Seconds (10.0));
```

ns-3 UDP client-server applications are suitable for simulating constant data rate traffic flows in a unidirectional manner. Suppose we want to confirm whether the server received all packets sent by a client application; then, we need to use ns-3 UDP echo client and server applications. We'll discuss these in the next section.

## ns-3 UDP echo client and server applications

In contrast to ns-3 UDP client-server applications, ns-3 UDP echo client-server applications exchange packets in both directions, that is, exchange echo requests and responses. That means the UDP echo server responds to every packet received from the connected UDP clients. General ns-3 UDP echo client and server application packet exchange is shown in *Figure 3.9*:

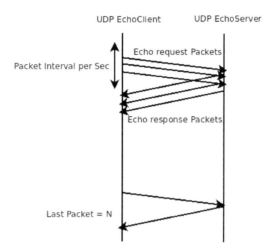

Figure 3.9 – The ns-3 UDP echo client-server traffic flow exchange

Let's see an example of UDP echo client and server application configuration in an ns-3 code snippet. First, create a UDP echo server application with a port number using the UdpEchoServerHelper class:

```
uint64_t port = 9;
UdpEchoServerHelper echoServer (port);
```

Next, install the UDP echo server application on a node with index-1. Configure suitable start and stop timings for the UDP echo server:

```
ApplicationContainer serverApps = echoServer.Install (nodes.
Get (1));
serverApps.Start (Seconds (1.0));
serverApps.Stop (Seconds (10.0));
```

After the UDP echo server is set up, create a UDP echo client application with the UDP echo server's IP address and port number as parameters. Configure its traffic characteristics such as MaxPackets, Interval, and PacketSize:

```
UdpEchoClientHelper echoClient (interfaces.GetAddress (1),
port);
  echoClient.SetAttribute ("MaxPackets", UintegerValue
(100000));
  echoClient.SetAttribute ("Interval", TimeValue (Seconds
(0.001)));
  echoClient.SetAttribute ("PacketSize", UintegerValue (1024));
```

Finally, install the UDP echo client on a node with index 0 and configure suitable start and stop timings to send all its packets:

```
ApplicationContainer clientApps = echoClient.Install (nodes.
Get (0));
  clientApps.Start (Seconds (2.0));
  clientApps.Stop (Seconds (10.0));
```

By receiving the UDP client application's echo request packets, the UDP echo server just echoes its received packets as echo replies to the respective client. UDP echo client-server applications are useful to simulate and test whether constant UDP data rate transmission and reception between two network nodes are successful or not.

## ns-3 TCP bulk send and packet sink applications

ns-3 offers the TCP BulkSendHelper class to simulate TCP traffic flow generation. Users can install BulkSendHelper on an ns-3 node and configure it to send N MaxBytes. Then, the TCP bulk send application's installed node sends a number of TCP segments by following all TCP protocol rules. To collect TCP data on the destination node, you should install the PacketSink application on the particular node in a simulation setup. In a simulation, the PacketSink application does not send any explicit TCP application data, but the PacketSink application's installed node's TCP protocol sends necessary TCP ACKs to the source node's TCP protocol. The working of the TCP BulkSender application and PacketSink application is shown in *Figure 3.10*:

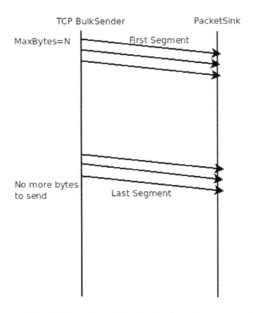

Figure 3.10 – The ns-3 TCP BulkSender and PacketSink applications traffic flow exchange

Let's see how to use the ns-3 `BulkSendHelper` and `PacketSink` applications to simulate the following scenario:

- In a simulation topology from node-0, the BulkSend application wants to send 300,000 bytes to node-1
- The packet sink application is installed on node-1

First, create a TCP data sending application using the ns-3 `BulkSendHelper` class with node-1's destination IP and port number:

```
uint64_t port = 9;
BulkSendHelper sourceHelper ("ns3::TcpSocketFactory",
                              InetSocketAddress (interfaces.
GetAddress (1), port));
```

Then, configure the `MaxBytes` parameter of the bulk sender application to set the maximum transmission data size. If you set `MaxBytes` to 0, it means the TCP `BulkSender` application sends TCP segments continuously until the simulator stops. The default value of `MaxBytes` is 0:

```
sourceHelper.SetAttribute ("MaxBytes", UintegerValue
(300000));
```

Next, install the bulk sender application on node-0 using `ApplicationContainer` and configure its suitable start and stop timings to send `MaxBytes` completely:

```
ApplicationContainer sourceApp = sourceHelper.Install (nodes.
Get (0));
sourceApp.Start (Seconds (0.0));
sourceApp.Stop (Seconds (10.0));
```

Finally, install the packet sink application (ns-3's `PacketSinkHelper`) on node-1 using `ApplicationContainer` and configure suitable start and stop timings for it to receive `MaxBytes` completely:

```
PacketSinkHelper sinkHelper ("ns3::TcpSocketFactory",
                             InetSocketAddress (Ipv4Address::GetAny
(), port));
ApplicationContainer sinkApp = sinkHelper.Install (nodes.Get
(1));
sinkApp.Start (Seconds (0.0));
sinkApp.Stop (Seconds (12.0));
```

It is that simple to configure TCP application flows in an ns-3 simulation using `BulkSendHelper` and `PacketSink` applications. This helps you to simulate TCP file transfer applications between two nodes.

## ns-3 On-Off TCP/UDP applications

ns-3 offers `OnOffApplication` to simulate interesting TCP or UDP traffic flows such as constant data rate or variable data rate applications. Users can create an `OnOffApplication` object using either TCP or UDP sockets and configure its `OnTime` and `OffTime` parameters to simulate constant data rate or variable data rate traffic flows. During `OnTime`, only packet exchange is done between nodes. During `OffTime`, no packet exchange is done. Moreover, to receive `OnOffApplication`-generated data on the destination node, users need to install the `PacketSink` application. ns-3's `OnOffApplication` and `PacketSink` application traffic flow exchange is shown in *Figure 3.11*:

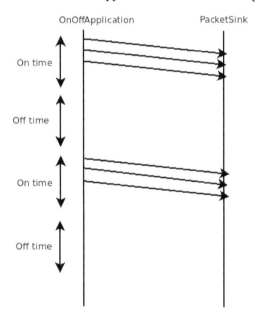

Figure 3.11 – The OnOffApplication and PacketSink traffic flow exchange

Let's see how to create an ns-3 TCP `OnOffApplication` and how to use it in our simulation using the following code snippets.

Start with creating a TCP `PacketSink` application using the ns-3 `PacketSinkHelper` class and install it on a node with index-1. Configure it with suitable start and stop timings. Hence, it receives all TCP data generated by `OnOffApplication`:

```
uint64_t port = 9;
Address localAddress (InetSocketAddress (Ipv4Address::GetAny
```

```
(), port));
  PacketSinkHelper sink ("ns3::TcpSocketFactory",
localAddress);
  ApplicationContainer apps_sink = sink.Install (nodes.Get
(1));
  apps_sink.Start (Seconds (0.5));
  apps_sink.Stop (Seconds (10));
```

While setting up applications, it can be noted that as we are interested in creating a TCP On-Off application, a TCPSocketFactory argument is passed to the PacketSinkHelper application object. If UDP OnOffApplication data collection is needed, we can configure UdpSocketFactory with the PacketSinkHelper object.

Next, we'll learn how to create and configure OnOffApplication using ns-3's OnOffHelper class.

First, we create OnOffApplication with TCPSocketFactor, the PacketSink node's IP, and port number arguments:

```
OnOffHelper onoff ("ns3::TcpSocketFactory", InetSocketAddress
(interfaces.GetAddress(1), port));
```

While creating the application, it can be noted that as we are interested in creating a TCP On-Off application, the TCPSocketFactory argument is passed to the OnOffHelper application object. We can configure UDP OnOffApplication with a UdpSocketFactory argument.

Now, let's see how to simulate the following traffic flow using the following code snippet:

- For example, we want our TCP application to transmit its data at a constant rate of 10 Mbps and each packet size is 1,000 bytes.

- We also want our TCP application to be off for every alternate second. That means after the application starts sending its data for 1 second, for the next 1 second it should be stopped. That means OnOffApplication OnTime is 1 second and OffTime is also 1 second. This process should be repeated until OnOffApplication stops:

```
onoff.SetConstantRate (DataRate ("10Mb/s"), 1000);
onoff.SetAttribute ("OnTime",  StringValue
("ns3::ConstantRandomVariable[Constant=1]"));
onoff.SetAttribute ("OffTime",  StringValue
("ns3::ConstantRandomVariable[Constant=1]"));
```

If the application wants to send data continuously at a constant rate, OffTime should be configured to 0.

Finally, configure OnOffApplication with suitable start and stop timings and install it on a node with index 0:

```
onoff.SetAttribute ("StartTime", TimeValue (Seconds (0.5)));
onoff.SetAttribute ("StopTime", TimeValue (Seconds (10)));
ApplicationContainer apps_source = onoff.Install (nodes.Get
(0));
```

Next, let's learn how to simulate HTTP applications using ns-3.

## ns-3 HTTP client and server applications

ns-3 offers HTTP client and server applications to simulate website browsing scenarios. To implement and simulate various browsing scenarios in your simulation setup, the ns-3 ThreeGppHttpServerHelper and ThreeGppHttpClientHelper classes are helpful. ns-3's 3GPP HTTP server implementation simulates generating various HTTP objects, such as Main and Embedded objects of different sizes in terms of bytes. Actually, these objects are generated based on a random distribution with mean and standard deviation values. Users can configure main and embedded objects' sizes in terms of mean and standard deviation. On the other hand, the ns-3 ThreeGppHttpClient application received these objects in a number of HTTP response packets based on the **Maximum Transmission Unit (MTU)** of its TCP socket used in the application. We can understand how the ns-3 HTTP server and client applications exchange main and embedded objects by looking at *Figure 3.12*:

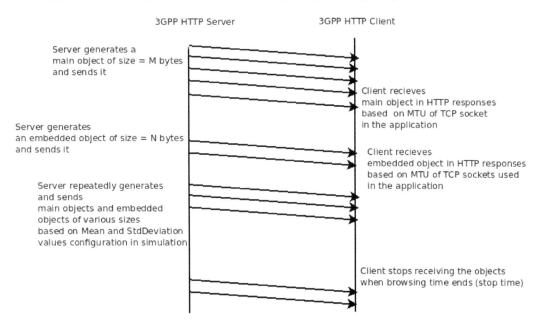

Figure 3.12 – The ns-3 HttpServer and HttpClient applications' traffic flow exchange

Let's see how to use ns-3 `ThreeGppHttpServer` and client applications to simulate a browsing scenario by setting the server to generate the main object's mean and standard deviation values.

First, create and configure the 3GPP HTTP server using the ns-3 `ThreeGppHttpServerHelper` and `ThreeGppHttpVariables` classes with the following code snippet.

Let's start with installing the HTTP server on the node with index-1 by configuring the server's IP address:

```
Ipv4Address serverAddress = interfaces.GetAddress (1);
 ThreeGppHttpServerHelper serverHelper (serverAddress);
 ApplicationContainer serverApps = serverHelper.Install
(nodes.Get (1));
```

Then, configure the main objects' mean and standard deviation values using the `ThreeGppHttpServer` class and its `Variables` attribute:

```
Ptr<ThreeGppHttpServer> httpServer = serverApps.Get
(0)->GetObject<ThreeGppHttpServer> ();
  PointerValue varPtr;
 httpServer->GetAttribute ("Variables", varPtr);
 Ptr<ThreeGppHttpVariables> httpVariables = varPtr.
Get<ThreeGppHttpVariables> ();
 httpVariables->SetMainObjectSizeMean (10240);
 httpVariables->SetMainObjectSizeStdDev (40960)
```

Finally, create and install an HTTP client application on the node with index 0 using the ns-3 `ThreeGppHttpClientHelper` class, then set the end time of the HTTP client's browsing time by configuring the client's stop time for closing the browsing session between the HTTP client and HTTP server applications:

```
ThreeGppHttpClientHelper clientHelper (serverAddress);
 ApplicationContainer clientApps = clientHelper.Install
(nodes.Get (0));
 clientApps.Stop (Seconds (1000));
```

Well done! We have successfully explored various ns-3-supported network applications and how to use these applications in our simulations. Next, we will revise all the topics explored in this chapter until now by setting up a simple LAN and Wi-Fi LAN using step-by-step procedures.

# The step-by-step procedure for building a simple LAN

In this section, we are going to learn how to set up a CSM LAN using a 24-port switch/bridge in ns-3. Let's see how to set up the LAN with the following specifications:

- The maximum number of hosts is 24
- The CSMA link speed is 1 Gbps and the propagation delay is 1 nanoseconds
- Use a 24-port switch and connect all hosts to it using CSMA links
- Test connectivity between nodes using the ns-3 ping application
- Check the LAN speed using ns-3 `OnOffApplication`

Let's write the following ns-3 code snippets in `pkt_simplelan.cc` to simulate the CSMA LAN.

Start by including all necessary ns-3 modules:

```
#include "ns3/core-module.h"
#include "ns3/network-module.h"
#include "ns3/bridge-module.h"
#include "ns3/internet-apps-module.h"
#include "ns3/csma-module.h"
#include "ns3/internet-module.h"
#include "ns3/applications-module.h"
#include "ns3/flow-monitor-module.h"
using namespace ns3;
```

In the `main()` simulation program, create 24 ns-3 nodes for 24 hosts, and 1 more ns-3 node for configuring the switch or bridge:

```
int
main (int argc, char *argv[])
{
  NodeContainer lan1;
  lan1.Create(24);
  Ptr<Node> bridge1 = CreateObject<Node> ();
```

After creating the necessary nodes, create and configure the `CsmaChannel` speed and delay using the ns-3 `CsmaHelper` class:

```
  CsmaHelper csma;
  csma.SetChannelAttribute ("DataRate", StringValue ("1Gbps"));
  csma.SetChannelAttribute ("Delay", StringValue ("1ns"));
```

Next, connect all hosts to the bridge using CSMA channels and collect host network devices and bridge network devices separately for later configuration:

```
NetDeviceContainer hostDevices;
NetDeviceContainer bridgeDevices;
for (int i = 0; i < 24; i++)
    {
       NetDeviceContainer link = csma.Install (NodeContainer
(lan1.Get (i), bridge1));
       hostDevices.Add (link.Get (0));
       bridgeDevices.Add (link.Get (1));
    }
```

Now, we can install bridge protocols and bridge devices on the ns-3 bridge node using the ns-3 BridgeHelper class:

```
BridgeHelper bridge;
bridge.Install (bridge1, bridgeDevices);
```

Next, install the internet stack and configure IP addresses to hosts using the ns-3 InternetStackHelper class:

```
InternetStackHelper internet;
internet.Install (lan1);
Ipv4AddressHelper ipv4;
ipv4.SetBase ("192.168.1.0", "255.255.255.0");
Ipv4InterfaceContainer interfaces = ipv4.Assign (lan1devices);
```

Till this point, we have created the CSMA LAN by connecting 24 hosts to the bridge using CSMA channels. Next, we need to test LAN connectivity and LAN link speed.

First, we show how to test connectivity among hosts using the ping application (ns-3 V4PingHelper). In our testing, host-0 pings to all other hosts and confirms the connectivity using the following code snippet:

```
for (int i=0;i<24;i++)
{
   ApplicationContainer p;
   V4PingHelper ping = V4PingHelper (interfaces.GetAddress
(i));
   ping.SetAttribute("Verbose",StringValue("true"));
```

```
    NodeContainer pingers;
    pingers.Add (lan1.Get (0));
    p = ping.Install (pingers);
    p.Start (Seconds (2.0));
    p.Stop (Seconds (5.0));
}
```

Finally, we will show you how to test LAN link speed by installing OnOffApplication on host-0 using the ns-3 OnOffHelper class and the packet sink application on host-23 using the ns-3 PacketSinkHelper class.

First, create a packet sink application and configure its start and stop timings, then install it on host-23 of the LAN:

```
    uint64_t port = 9;
    Address localAddress (InetSocketAddress (Ipv4Address::GetAny
    (), port));
    PacketSinkHelper sink ("ns3::UdpSocketFactory",
    localAddress);
    ApplicationContainer apps_sink = sink.Install (lan1.Get
    (23));
    apps_sink.Start (Seconds (0.5));
    apps_sink.Stop (Seconds (10));
```

Next, we can install either a TCP or UDP application with a constant data rate of 1 Gbps sending the OnOffApplication on host-0 to the packet sink application of host-23:

```
    OnOffHelper onoff ("ns3::UdpSocketFactory", InetSocketAddress
    (interfaces.GetAddress(23), port));
    onoff.SetConstantRate (DataRate ("1Gbps"), 1000);
```

To test the link speed, we start the UDP application after the ping application ends (at 6 seconds) and stop it after 2 seconds (at 8 seconds). The OnOffApplication parameters are configured as follows:

```
    onoff.SetAttribute ("StartTime", TimeValue (Seconds (6)));
    onoff.SetAttribute ("OnTime",  StringValue
    ("ns3::ConstantRandomVariable[Constant=1]"));
    onoff.SetAttribute ("OffTime", StringValue
    ("ns3::ConstantRandomVariable[Constant=0]"));
    onoff.SetAttribute ("StopTime", TimeValue (Seconds (8)));
    ApplicationContainer apps_source = onoff.Install (lan1.Get
    (0));
```

Let's close the main program by installing Flow Monitor and collecting results into the `lanspeed.xml` file:

```
FlowMonitorHelper flowmonHelper;
flowmon = flowmonHelper.InstallAll ();
Simulator::Stop (Seconds(11.0));
Simulator::Run ();
flowmon->SerializeToXmlFile ("lanspeed.xml", true, true);
Simulator::Destroy ();
}
```

That's great. We have successfully set up a CSMA LAN of 24 hosts using a bridge. Let's test it by running the simulation (pkt_simplelan.cc) from our ns-3 installed folder (ns-allinone-3.36/ns-3.36). It collects results in the ping and lanspeed.xml files:

```
./ns3 run scratch/pkt_simplelan | tee ping
```

First, open the ping file to see the ping results. It contains each host ping result saying **3 packets transmitted, 3 packets received, 0% packet loss.**. So, we can confirm that all other hosts are reachable from host-0.

You can check each host ping result using the following command by passing the respective host IP address in the grep command argument:

```
cat ping |grep '192.168.1.4'
```

It displays the ping results from host-0 to the host with the IP 192.168.1.4, as shown in *Figure 3.13*:

```
PING 192.168.1.4 - 56 bytes of data - 84 bytes including ICMP and IPv4 headers.
64 bytes from 192.168.1.4: icmp_seq=0 ttl=64 time=+6.01732ms
64 bytes from 192.168.1.4: icmp_seq=1 ttl=64 time=+0.008092ms
64 bytes from 192.168.1.4: icmp_seq=2 ttl=64 time=+0.052092ms
```

Figure 3.13 – The ping results from Host-0 to host IP 192.168.1.4

We can also check the **Round Trip Time (RTT)** results using the following command:

```
cat ping |grep 'rtt'
```

Our LAN setup RTT results from the `ping` file are displayed in *Figure 3.14*:

```
rtt min/avg/max/mdev = 0/2.667/8/4.619 ms
rtt min/avg/max/mdev = 0/4.667/14/8.083 ms
rtt min/avg/max/mdev = 0/1.333/4/2.309 ms
rtt min/avg/max/mdev = 0/4.667/14/8.083 ms
rtt min/avg/max/mdev = 0/4.333/13/7.506 ms
```

Figure 3.14 – Our LAN setup RTT results captured in a ping file

Now, to check the LAN link speed results, let's parse the `lanspeed.xml` file using the following command:

```
python3 scratch/flowmon-parse-results.py lanspeed.xml
```

Our LAN setup link speed from flow results file is displayed in *Figure 3.15*.

```
FlowID: 1 (UDP 192.168.1.1/49153 --> 192.168.1.24/9)
         TX bitrate: 1028006.17 kbit/s
         RX bitrate: 971645.76 kbit/s
         Mean Delay: 53.93 ms
         Packet Loss Ratio: 1.35 %
```

Figure 3.15 – Our simulation LAN setup link speed results

From the results, we can observe that host-0 (`192.168.1.1`) is generating 1 Gbps data to host-23 (`192.168.1.23`). Host-23 is able to receive 971 Mbps successfully. Hence, we can confirm the LAN is working at the configured speed.

> **Important note**
>
> As the CSMA channel is a shared channel, in the case of the simultaneous access of multiple hosts, the link speed will be shared among them.

It is that simple to set up and evaluate a simple wired LAN in ns-3. Next, we will do a hands-on task related to setting up a simple Wi-Fi LAN.

# The step-by-step procedure for building a simple wireless LAN

In this section, we are going to learn how to set up a simple wireless LAN with 1 Wi-Fi AP and 20 Wi-Fi STAs by doing the following:

1.  Set the Wi-Fi standard as 802.11ac.
2.  Set up a Wi-Fi AP with the SSID `Packt`.

3.  Set up 20 Wi-Fi STAs.

4.  Randomly place the 20 Wi-Fi STAs around the AP within a radius of 30 meters.

5.  Allow all STAs to move around the AP.

6.  Test the connectivity between all STAs and the AP.

7.  Test the link speed between an STA and the AP.

Let's write the following code snippets in `simple_wifi_lan.cc` to simulate a simple wireless LAN.

First import all the necessary ns-3 packages:

```
#include "ns3/core-module.h"
#include "ns3/yans-wifi-helper.h"
#include "ns3/yans-wifi-channel.h"
#include "ns3/ssid.h"
#include "ns3/mobility-helper.h"
#include "ns3/internet-module.h"
#include "ns3/internet-apps-module.h"
#include "ns3/ipv4-address-helper.h"
#include "ns3/applications-module.h"
#include "ns3/flow-monitor-module.h"
using namespace ns3;
```

At the start of `main ()`, create the necessary ns-3 nodes for configuring APs and STAs:

```
int main (int argc, char *argv[])
{
  uint32_t nAPs = 1;
  uint32_t nStas = 20;
  NodeContainer APs;
  APs.Create (nAPs);
  NodeContainer stas;
  sta.Create (nStas);
```

Configure and install the Wi-Fi AP placement model using the ns-3 `ConstantPositionMobilityModel` class at the default position (0,0):

```
  MobilityHelper mobility;
  Ptr<ListPositionAllocator> positionAlloc =
CreateObject<ListPositionAllocator> ();
```

```
    positionAlloc->Add (Vector (100, 100, 0.0));
    mobility.SetPositionAllocator (positionAlloc);
    mobility.SetMobilityModel
  ("ns3::ConstantPositionMobilityModel");
    mobility.Install (APs.Get (0));
```

Next, randomly place STAs around the AP using the ns-3 RandomDiscPositionAllocator class. You need to configure a random disc center with X and Y coordinates, and the maximum radius from the disc center using the Rho variable for the random placement of STAs:

```
  mobility.SetPositionAllocator
  ("ns3::RandomDiscPositionAllocator",
                              "X", StringValue ("100.0"),
                              "Y", StringValue ("100.0"),
                              "Rho", StringValue
  ("ns3::UniformRandomVariable[Min=0|Max=30]"));
```

Now, configure the STAs' mobility around the AP using the ns-3 RandomWalk2dMobilityModel class with the following code snippets. The mobility speed and the boundaries of the mobility region need to be configured. Here, Bounds' Max X and Max Y values should be greater than PositionAllocator's Max X and Max Y values:

```
  mobility.SetMobilityModel ("ns3::RandomWalk2dMobilityModel",
                              "Mode", StringValue ("Time"),
                              "Time", StringValue ("100s"),
                              "Speed", StringValue
  ("ns3::ConstantRandomVariable[Constant=2.0]"),
                              "Bounds", StringValue
  ("0|200|0|200"));
    mobility.Install (sta);
```

Now, the wireless LAN node placement and mobility configurations are done. Next, set the Wi-Fi standard as 802.11ac and create a standard common Wi-Fi PHY channel for both the AP and STAs using ns-3's YansWifiPhyHelper and YansWifiChannelHelper classes:

```
  NetDeviceContainer staDev;
  NetDeviceContainer apDev;
  WifiHelper wifi;
  wifi.SetStandard (WIFI_STANDARD_80211ac);
  YansWifiPhyHelper wifiPhy;
```

```
wifiPhy.SetPcapDataLinkType (WifiPhyHelper::DLT_IEEE802_11_
RADIO);
   YansWifiChannelHelper wifiChannel =
YansWifiChannelHelper::Default ();
   wifiPhy.SetChannel (wifiChannel.Create ());
```

After configuring the Wi-Fi PHY channel, set up the AP by configuring its Wi-Fi MAC with the SSID Packt using the following code snippet:

```
WifiMacHelper wifiMac;
Ssid ssid = Ssid ("Packt");
wifiMac.SetType ("ns3::ApWifiMac",
                   "Ssid", SsidValue (ssid));
```

Then, install the Wi-Fi PHY and MAC layers on the ns-3 AP node:

```
apDev = wifi.Install (wifiPhy, wifiMac, APs.Get (0));
```

Similar to with APs, now configure the STA-specific Wi-Fi MAC and install the PHY layer and MAC layer on ns-3 STA nodes using the following code snippet:

```
wifiMac.SetType ("ns3::StaWifiMac",
                   "Ssid", SsidValue (ssid));
staDev = wifi.Install (wifiPhy, wifiMac, sta);
```

Now, we can install the internet stack and configure IP addresses to the AP and STAs using the following code snippet:

```
InternetStackHelper stack;
stack.Install(APs);
stack.Install (sta);
Ipv4AddressHelper ip;
Ipv4InterfaceContainer staInterface;
Ipv4InterfaceContainer apInterface;
ip.SetBase ("192.168.0.0", "255.255.255.0");
    // assign AP IP address to bridge, not wifi
apInterface = ip.Assign (apDev);
staInterface = ip.Assign (staDev);
```

We have successfully set up the wireless LAN topology node placement and mobility models. The AP and STAs are installed with suitable PHY, MAC, and TCP/IP stack protocols. Next, let's start installing the necessary application for testing connectivity between all STAs and the AP. Like LAN connectivity testing, the AP is also configured here with an ns-3 ping application to test the connectivity between the AP and all its STAs using the following code snippet:

```
for (int i=0;i<nStas;i++)
{
    ApplicationContainer p;
    V4PingHelper ping = V4PingHelper (staInterface.GetAddress
(i));
    ping.SetAttribute("Verbose",StringValue("true"));
    NodeContainer pingers;
    pingers.Add (APs.Get (0));
    p = ping.Install (pingers);
    p.Start (Seconds (2.0));
    p.Stop (Seconds (5.0));
}
```

Finally, to test the Wi-Fi link speed between an STA and AP, we use UDP's OnOffApplication with constant data rate generation. To test the link speed, we start the on-off UDP application after the ping application ends (at 6 seconds) and stop it after 2 seconds (at 8 seconds). In the following code snippet, UDP's OnOffApplication generates 100 Mbps traffic at STA-1 to the packet sink application configured at the AP:

> **Important note**
> If 100 Mbps is leading to high packet loss, you may decrease the DataRate value; if there is no packet loss, try a higher DataRate configuration.

```
OnOffHelper onoff ("ns3::UdpSocketFactory", InetSocketAddress
(apInterface.GetAddress(0), port));
  onoff.SetConstantRate (DataRate ("100Mbps"), 1000);
  onoff.SetAttribute ("StartTime", TimeValue (Seconds (6)));
  onoff.SetAttribute ("OnTime",  StringValue
("ns3::ConstantRandomVariable[Constant=1]"));
  onoff.SetAttribute ("OffTime", StringValue
("ns3::ConstantRandomVariable[Constant=0]"));
  onoff.SetAttribute ("StopTime", TimeValue (Seconds (8)));
  ApplicationContainer apps_source = onoff.Install (sta.Get
(1));
```

To receive OnOffApplication-generated data at the AP, a packet sink application is installed on the ns-3 AP node using the following code snippet:

```
uint64_t port = 9;
Address localAddress (InetSocketAddress (Ipv4Address::GetAny
(), port));
PacketSinkHelper sink ("ns3::UdpSocketFactory",
localAddress);
ApplicationContainer apps_sink = sink.Install (APs.Get (0));
apps_sink.Start (Seconds (0.5));
apps_sink.Stop (Seconds (10));
```

Let's close the main program by installing Flow Monitor and collecting results into the wlanspeed. xml file:

```
Ptr<FlowMonitor> flowmon;
FlowMonitorHelper flowmonHelper;
flowmon = flowmonHelper.InstallAll ();
Simulator::Stop (Seconds(11.0));
Simulator::Run ();
flowmon->SerializeToXmlFile ("wlanspeed.xml", true, true);
Simulator::Destroy ();
}
```

Well done! We have learned how to simulate a simple wireless LAN and test it by writing simple_wifi_lan.cc. Let's test simple_wifi_lan.cc from the ns-3 installed directory (ns-allinone-3.36/ns-3.36). It collects results in the wping and wlanspeed.xml files:

**./ns3 run scratch/simple_wifi_lan | tee wping**

First, open the wping file to see the ping results. It contains each STA ping result, saying **3 packets transmitted, 3 packets received, 0% packet loss.**. That means we can confirm that all STAs are reachable from the AP.

You may check specific STA ping results using the following command by passing the respective STA IP address in the grep command argument:

**cat wping |grep '192.168.1.10'**

The ping results from the AP to STA `192.168.1.10` are displayed in *Figure 3.16*:

```
PING 192.168.1.10 - 56 bytes of data - 84 bytes including ICMP and IPv4 headers.
64 bytes from 192.168.1.10: icmp_seq=0 ttl=64 time=+39.9295ms
64 bytes from 192.168.1.10: icmp_seq=1 ttl=64 time=+3.79503ms
64 bytes from 192.168.1.10: icmp_seq=2 ttl=64 time=+3.85775ms
```

Figure 3.16 – The ping results from AP to STA 192.168.1.10

We can also check the RTT results using the following command:

```
cat wping |grep 'rtt'
```

The RTT results of our Wi-Fi LAN setup from the `wping` file are displayed in *Figure 3.17*:

```
rtt min/avg/max/mdev = 6/10.33/19/7.506 ms
rtt min/avg/max/mdev = 7/16/34/15.59 ms
rtt min/avg/max/mdev = 7/18.33/41/19.63 ms
rtt min/avg/max/mdev = 7/13/24/9.539 ms
```

Figure 3.17 – Our Wi-Fi LAN setup RTT results captured in the wping file

Now, to check the Wi-Fi link speed between STA-1 and the AP, parse the `wlanspeed.xml` file using the following command:

```
python3 scratch/flowmon-parse-results.py wlanspeed.xml
```

Our Wi-Fi setup link speed from the flow results is displayed in *Figure 3.18*:

```
FlowID: 1 (UDP 192.168.1.3/49153 --> 192.168.1.1/9)
        TX bitrate: 102804.11 kbit/s
        RX bitrate: 102721.41 kbit/s
        Mean Delay: 1.72 ms
        Packet Loss Ratio: 0.00 %
```

Figure 3.18 – Our Wi-Fi LAN setup link speed test results – 1

From the results, we can observe that STA-1 (`192.168.1.3`) is generating 100 Mbps data to the AP (`192.168.1.1`). The AP is able to receive 100 Mbps successfully. Let's change `DataRate` to 200 Mbps and check the following results in *Figure 3.19*:

```
FlowID: 1 (UDP 192.168.1.3/49153 --> 192.168.1.1/9)
        TX bitrate: 205604.11 kbit/s
        RX bitrate: 111549.41 kbit/s
        Mean Delay: 541.21 ms
        Packet Loss Ratio: 24.46 %
```

Figure 3.19 – Our Wi-Fi LAN setup link speed test results – 2

We can observe that the AP is able to receive only 110 Mbps and there is a packet loss of 24%. Hence, we can conclude our simple wireless LAN speed is working at around 100 Mbps.

> **Important note**
>
> The Wi-Fi network speed depends on the standard, such as 802.11/a/b/g/n/ac/ax. Also, as the Wi-Fi channel is a shared channel, if multiple STAs are accessing it, the link speed will be shared among them.

## Summary

In this chapter, you learned and practiced how to set up a variety of wired and wireless networking devices using key ns-3 building blocks. As part of this chapter, you also learned how to set up wired and/or wireless network topologies systematically using various ns-3-supported communication channels, such as CSMA, P2P, and Wi-Fi. This will help you to easily set up any complex wired or wireless network topologies. In addition, to evaluate network performance thoroughly and quickly, you were introduced to real-time network applications, such as ping, HTTP, and constant or variable data rate traffic generators.

Finally, to make you comfortable with setting up wired and wireless network topologies, relevant hands-on tasks were discussed, broken down into step-by-step procedures. These hands-on activities especially focused on how to deploy and test wired and Wi-Fi LAN link connectivity among all nodes, and link speed.

In the next chapter, we are going to cover interesting hands-on activities related to setting up and evaluating a variety of wired, Wi-Fi, heterogeneous, and internet topologies using ns-3.

# Part 2:
# Learn, Set Up, and Evaluate Wired and Wi-Fi (802.11a/b/g/n/ac/ax) Networks

On completion of *Part 2*, you will have learned how to install, configure, and test important network nodes and protocols of CSMA/P2P/Wi-Fi (802.11n/ac/ax) and ad hoc networks using ns-3. As part of your Wi-Fi research activities, you will learn how to set up and simulate ad hoc routing, and about 802.11n/ac/ax advanced features, including frame aggregation, channel bonding, MIMO, resource scheduling, and BSS coloring using ns-3-supporting Wi-Fi algorithms.

This part has the following chapters:

- *Chapter 4, Setting Up and Evaluating CSMA/P2P LANs, Wi-Fi LANs, and the Internet*
- *Chapter 5, Exploring Basic Wi-Fi Technologies and Setting Up and Evaluating Wireless Ad Hoc Networks*
- *Chapter 6, Researching Advanced Wi-Fi Technologies 802.11n, ac, and ax in ns-3*

# 4

# Setting Up and Evaluating CSMA/P2P LANs, Wi-Fi LANs, and the Internet

Having learned about the concepts of ns-3 simulation key building blocks and their basic features, you can start working on setting up a variety of network topologies and evaluating their performance systematically. In this chapter, you will learn how to set up and evaluate a wide variety of networks such as LANs, Wi-Fi LANs, heterogeneous LANs, and the internet. As part of these activities, we will discuss how to set up a topology and how to configure necessary protocols on a variety of hosts and suitable interconnecting networking devices.

In this chapter, you will learn how to test connectivity and link speed using suitable network applications on your topologies to evaluate network performance. Specifically, in order to simulate a real-time wireless LAN environment, you will be introduced to the necessary placement and mobility models. Besides using NetAnimator (as discussed in *Chapter 1, Getting Started with Network Simulator-3 (ns-3)*), we will discuss important details of wireless network simulations, such as how various Wi-Fi stations (STAs) are deployed, their mobility patterns, and various types of packets exchanged. In order to introduce setting up a wide variety of networks in a logical order, we will discuss the following hands-on activities with step-by-step procedures:

- Step-by-step procedure for setting up CSMA/CD LANs
- Step-by-step procedure for setting up Wi-Fi LANs
- Step-by-step procedure for setting up heterogeneous LANs (CSMA/CD and Wi-Fi LANs)
- Step-by-step procedure for setting up and evaluating internet simulation topology

# Step-by-step procedure for setting up CSMA/CD LANs

We have already learned how to set up a simple LAN in *Chapter 3*, *ns-3 Key Building Blocks for Simulations*, which helps to connect hosts in a network and exchange their data traffic. Now, we are going to learn a more realistic LAN setup (refer to *Figure 4.1*), which is generally deployed in educational institutes, offices, and common workspaces:

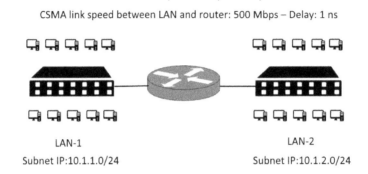

CSMA LANs link speed: 1 Gbps – Delay: 1 ns

CSMA link speed between LAN and router: 500 Mbps – Delay: 1 ns

LAN-1
Subnet IP:10.1.1.0/24

LAN-2
Subnet IP:10.1.2.0/24

Figure 4.1 – A realistic CSMA/CD LAN setup

In this hands-on activity, we set up CSMA/CD LANs with the following objectives:

- Set up individual LANs for specific workplaces such as labs, departments, and so on
- Connect each LAN using CSMA links of link speed: 1 Gbps and delay: 1 ns
- Interconnect LAN-1 and LAN-2 using a router using CSMA link speed: 500 Mbps and delay: 1 ns
- Configure separate subnet IP addresses for each LAN to limit broadcast traffic
- Test link speed within a LAN
- Test link speed between LANs
- Demonstrate how to share a file between LANs

## CSMA/CD LANs simulation in ns-3

Let's set up CSMA LANs simulation in a step-by-step procedure with the following ns-3 code snippets in pkt_csma_lans.c:.

1.  First, import all the following necessary ns-3 packages:

    ```
    #include "ns3/core-module.h"
    #include "ns3/network-module.h"
    ```

```
#include "ns3/applications-module.h"
#include "ns3/internet-apps-module.h"
#include "ns3/bridge-module.h"
#include "ns3/csma-module.h"
#include "ns3/internet-module.h"
#include "ns3/flow-monitor-module.h"
using namespace ns3;
```

2.  In main(), start with creating the necessary number of ns-3 nodes for setting up LAN-1 and LAN-2 hosts, routers, and switches using the respective ns-3 node containers. Note that in LAN-1 and LAN-2, node-23 will be configured for router setup:

```
NodeContainer lan1;
lan1.Create(23);
NodeContainer lan2;
lan2.Create(23);
NodeContainer router;
router.Create(1);
lan1.Add(router);
Ptr<Node> bridge1 = CreateObject<Node> ();
Ptr<Node> bridge2 = CreateObject<Node> ();
```

3.  Next, create a CSMA channel with a link speed of 1 Gbps and a delay of 1 ns for connecting hosts in a LAN using ns-3 CsmaHelper:

```
CsmaHelper csma;
csma.SetChannelAttribute ("DataRate",
StringValue("1Gbps"));
csma.SetChannelAttribute ("Delay", StringValue("1ns"));
```

4.  Now, let's connect LAN-1 hosts with the bridge1 (switch) using the CSMA channel and collect host network devices and bridge network devices separately for later configurations using the following ns-3 code snippets:

```
NetDeviceContainer lan1Devices;
NetDeviceContainer lan1BridgeDevices;
for (int i = 0; i < 22; i++)
  {
    NetDeviceContainer link = csma.Install
(NodeContainer (lan1.Get (i), bridge1));
```

```
          lan1Devices.Add (link.Get (0));
          lan1BridgeDevices.Add (link.Get (1));
        }
```

5.  To interconnect LAN-1 with LAN-2, connect them using a router (node-23) using another CSMA channel with a link speed of 500 Mbps and a delay of 1 ns:

```
        CsmaHelper rcsma;
        rcsma.SetChannelAttribute ("DataRate",
    StringValue("0.5Gbps"));
        rcsma.SetChannelAttribute ("Delay", StringValue
    ("1ns"));
        NetDeviceContainer rlink1 = rcsma.Install
    (NodeContainer (lan1.Get (23), bridge1));
        lan1Devices.Add (rlink1.Get (0));
        lan1BridgeDevices.Add (rlink1.Get (1));
```

6.  Complete LAN-1 cabling by configuring its bridge1 with protocols and attaching bridge network devices:

```
        BridgeHelper bridge;
        bridge.Install (bridge1, lan1BridgeDevices);
```

7.  After completing the cabling of LAN-1, install the internet protocol stack on all hosts and routers:

```
        NodeContainer allnodes (lan1,lan2);
        InternetStackHelper internet;
        internet.Install (allnodes);
```

8.  Next, start the LAN-2 setup by connecting all its hosts with bridge2 (switch) using the CSMA channel with link speed 1 Gbps and delay of 1 ns using the following ns-3 code snippets:

```
        NetDeviceContainer lan2Devices;
        NetDeviceContainer lan2BridgeDevices;
        for (int i = 0; i < 22; i++)
          {
            NetDeviceContainer link = csma.Install
    (NodeContainer (lan2.Get (i), bridge2));
            lan2Devices.Add (link.Get (0));
            lan2BridgeDevices.Add (link.Get (1));
          }
```

9.  To interconnect LAN-2 with LAN-1, connect its switch with the router using the CSMA channel with a link speed of 500 Mbps and a delay of 1 ns:

```
lan2.Add(router);
NetDeviceContainer rlink2 = rcsma.Install (NodeContainer
(lan2.Get (23), bridge2));
lan2Devices.Add (rlink2.Get (0));
lan2BridgeDevices.Add (rlink2.Get (1));
```

10. Complete LAN-2 cabling by configuring its `bridge2` with protocols and attaching bridge network devices:

```
bridge.Install (bridge2, lan2BridgeDevices);
```

11. Let's configure IP addresses for all hosts and routers according to their LAN subnet IP addresses. Then, configure all necessary routing rules on all nodes using the following ns-3 code snippets:

```
Ipv4AddressHelper ipv4;
ipv4.SetBase ("10.1.1.0", "255.255.255.0");
Ipv4InterfaceContainer interfaces1= ipv4.Assign
(lan1Devices);
ipv4.SetBase ("10.1.2.0", "255.255.255.0");
Ipv4InterfaceContainer interfaces2= ipv4.Assign
(lan2Devices);
Ipv4GlobalRoutingHelper::PopulateRoutingTables ();
```

That's great. We are done with successfully setting up LAN-1 and LAN-2 for their hosts' communication.

Now, we can proceed with the following ns-3 codes to test LAN-1 and LAN-2 thoroughly:

1.  Start with testing the LAN-1 link speed by generating a 1-Gbps constant data rate traffic between host-0 and host-23 using the following `OnOff` application setup code snippet:

```
OnOffHelper onoff ("ns3::UdpSocketFactory",
InetSocketAddress (interfaces1.GetAddress(22), port));
  onoff.SetConstantRate (DataRate ("1Gbps"), 1000);
  onoff.SetAttribute ("StartTime", TimeValue (Seconds
(2)));
  onoff.SetAttribute ("StopTime", TimeValue (Seconds
(4)));
  onoff.SetAttribute ("OnTime",  StringValue
("ns3::ConstantRandomVariable[Constant=1]"));
```

```
    onoff.SetAttribute ("OffTime", StringValue
("ns3::ConstantRandomVariable[Constant=0]"));
    ApplicationContainer apps_source1 = onoff.Install
(lan1.Get (0));
```

Similarly, we can test the LAN-2 link speed.

2.  Next, test the link speed between LAN-1's host-22 and LAN-2's host-0 using the following code snippet. This time, we configure the OnOff application after the LAN-1 link test is completed:

```
    onoff.SetAttribute ("StartTime", TimeValue (Seconds
(6)));
    onoff.SetAttribute ("StopTime", TimeValue (Seconds
(8)));
    ApplicationContainer apps_source2 = onoff.Install
(lan2.Get (0));
```

3.  Finally, we test a file sharing of size 10,240,000 bytes between LAN-1's host-0 and LAN-2's host-1 using the TcpBulkSend and PacketSink applications:

```
    BulkSendHelper sourceHelper ("ns3::TcpSocketFactory",
                                 InetSocketAddress
(interfaces2.GetAddress (1), port));
    sourceHelper.SetAttribute ("MaxBytes", UintegerValue
(10240000));
    ApplicationContainer sourceApp = sourceHelper.Install
(lan1.Get (0));
    sourceApp.Start (Seconds (10.0));
    sourceApp.Stop (Seconds (20.0));


    PacketSinkHelper sinkHelper ("ns3::TcpSocketFactory",
                                 InetSocketAddress
(Ipv4Address::GetAny (), port));
    ApplicationContainer sinkApp = sinkHelper.Install
(lan2.Get (1));
    sinkApp.Start (Seconds (9.0));
    sinkApp.Stop (Seconds (21.0));
```

4.  After writing the LAN's testing code, let's close main() by installing a flow monitor for performance evaluation and freeing all simulation resources using the following ns-3 code snippets:

```
    Simulator::Stop (Seconds(21.0));
    Simulator::Run ();
```

```
flowmon->SerializeToXmlFile ("csmalanspeed.xml", true,
true);
    Simulator::Destroy ();
    Simulator::Run ();
    Simulator::Destroy ();
}
```

That's great! We set up a CSMA LAN simulation successfully with a hands-on activity using ns-3. Let's evaluate our CSMA LAN simulation topology thoroughly in the next section.

## CSMA/CD LANs evaluation in ns-3

Let's run the pkt_csma_lans.cc simulation at your ns-3 installation directory and observe the results using the following command:

**./ns3 run scratch/pkt_csma_lans**

First, observe the LAN-1 link speed results by parsing csmalanspeed.xml using the following command at your ns-3 installation directory:

**python3 scratch/flowmon-parse-results.py csmalanspeed.xml**

We can observe that the LAN-1 link speed (shown in *Figure 4.2*) between (host-0 and host-22) is around 971 Mbps:

```
FlowID: 1 (UDP 10.1.1.1/49153 --> 10.1.1.24/9)
        TX bitrate: 1028006.17 kbit/s
        RX bitrate: 971648.94 kbit/s
        Mean Delay: 53.90 ms
        Packet Loss Ratio: 1.49 %
```

Figure 4.2 – LAN-1 link speed

It means that LAN-1 is working at a CSMA channel speed of 1 Gbps. Next, observe the link speed between LAN-1 and LAN-2 (shown in *Figure 4.3*):

```
FlowID: 2 (UDP 10.1.2.1/49153 --> 10.1.1.24/9)
        TX bitrate: 1028001.03 kbit/s
        RX bitrate: 485829.82 kbit/s
        Mean Delay: 55.62 ms
        Packet Loss Ratio: 50.82 %
```

Figure 4.3 – Link speed between LAN-1 and LAN-2

We observe that there is around 50% packet loss, and the link speed is around 485 Mbps only. This is due to the bottleneck link CSMA channel 500 Mbps, which was used to connect LAN-1 and LAN-2 using a router. Hence, the results are perfect.

Let's see the file transfer results (shown in *Figure 4.4*) between LAN-2 host-0 and LAN-1 host-1:

```
FlowID: 3 (TCP 10.1.1.1/49153 --> 10.1.2.2/9)
        TX bitrate: 60279.56 kbit/s
        RX bitrate: 60557.25 kbit/s
        Mean Delay: 0.46 ms
        Packet Loss Ratio: 0.14 %
FlowID: 4 (TCP 10.1.2.2/9 --> 10.1.1.1/49153)
        TX bitrate: 2859.84 kbit/s
        RX bitrate: 2798.54 kbit/s
        Mean Delay: 1.23 ms
        Packet Loss Ratio: 1.78 %
```

Figure 4.4 – File downloading speed between LAN-1 and LAN-2

From the results, we can observe that at around 60 Mbps speed, the file exchange is successfully done between LAN-1's host-0 and LAN-2's host-1.

Well done! We have completed the setup and thorough evaluation of CSMA LANs using ns-3. As part of the evaluations, we discussed how to test LAN link speed, the interworking of LANs, and identifying bottleneck links. Next, let's discuss how to set up and evaluate Wi-Fi LANs in the next section.

## Step-by-step procedure for setting up Wi-Fi LANs

In *Chapter 3*, we learned how to set up a simple Wi-Fi network. Now, we are going to learn how to set up a wireless LAN (refer to *Figure 4.5*) environment to support network access during STAs mobility within an area:

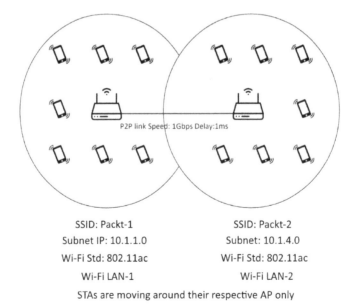

Figure 4.5 – Two Wi-Fi LANs connected using a dedicated P2P channel

For instance, in airports and shopping malls, operators deploy Wi-Fi APs to provide internet access for users. In this section, we will do a hands-on activity to simulate Wi-Fi network deployment in a real-time environment using ns-3 with the following objectives:

- For instance, we are assuming setting up two Wi-Fi LANs at airport shops
- Configure Wi-Fi standard as 802.11ac, and set `Packt-1` and `Packt-2` as **Service Set Identifiers (SSIDs)** to Wi-Fi LAN-1 and Wi-Fi LAN-2, respectively
- Configure each Wi-Fi LAN in its own IP subnet
- Users at two shops should communicate using their Wi-Fi LANs
- Two Wi-Fi LANs are connected using a dedicated P2P link speed of 1 Gbps and a delay of 1 ms
- Users are free to move within their shop boundaries only
- Test whether all Wi-Fi STAs in a shop area are reachable to their respective AP
- Test whether it is possible to communicate STAs between SSIDs
- Test Wi-Fi link speed between two STA
- Visualize Wi-Fi network simulation using NetAnimator

## Wi-Fi LANs simulation in ns-3

Let's set up Wi-Fi LANs simulation in a step-by-step procedure with the following ns-3 code snippets in `pkt_wifi_lans.c:`.

1. First, import all the following necessary ns-3 packages:

```
#include "ns3/core-module.h"
#include "ns3/point-to-point-module.h"
#include "ns3/network-module.h"
#include "ns3/applications-module.h"
#include "ns3/internet-apps-module.h"
#include "ns3/mobility-module.h"
#include "ns3/internet-module.h"
#include "ns3/yans-wifi-helper.h"
#include "ns3/ssid.h"
#include "ns3/flow-monitor-module.h"
#include "ns3/netanim-module.h"
using namespace ns3;
```

2.  In `main()`, start with creating the necessary number of ns-3 nodes for setting up Wi-Fi LAN-1's AP and STAs:

```
uint32_t nWifi = 10;
NodeContainer WiFiAPNodes;
WiFiAPNodes.Create (2);
NodeContainer wifiStaNodes1;
wifiStaNodes1.Create (nWifi);
```

3.  Next, create a dedicated P2P channel with a link speed of 1 Gbps and a delay of 1 ns for connecting two Wi-Fi APs in a LAN using the following ns-3 code snippet:

```
PointToPointHelper pointToPoint;
pointToPoint.SetDeviceAttribute ("DataRate",
StringValue ("1Gbps"));
pointToPoint.SetChannelAttribute ("Delay", StringValue
("1ms"));
NetDeviceContainer p2pDevices;
p2pDevices = pointToPoint.Install (WiFiAPNodes);
```

4.  Deploy the first Wi-Fi AP at a shop location (X,Y) (100,100) using the following code snippet:

```
NodeContainer wifiApNode1 = WiFiAPNodes.Get (0);
MobilityHelper mobility;
Ptr<ListPositionAllocator> positionAlloc =
CreateObject<ListPositionAllocator> ();
positionAlloc->Add (Vector (100, 100, 0.0));
mobility.SetPositionAllocator (positionAlloc);
mobility.SetMobilityModel
("ns3::ConstantPositionMobilityModel");
mobility.Install (wifiApNode1);
```

5.  After deploying the first Wi-Fi AP, place all its STAs at random locations (only within a 20-meter (m) radius from the AP) around the AP using the following ns-3 code snippet:

```
mobility.SetPositionAllocator
("ns3::RandomDiscPositionAllocator",
                            "X", StringValue ("100.0"),
                            "Y", StringValue ("100.0"),
                            "Rho", StringValue
("ns3::UniformRandomVariable[Min=0|Max=20]"));
```

6. Next, configure the random walk mobility model with max speed allowed (e.g., 4m/s), and boundaries for all its STAs using the following ns-3 code snippet:

```
mobility.SetMobilityModel
("ns3::RandomWalk2dMobilityModel",
                            "Mode", StringValue
("Time"),
                            "Time", StringValue
("100s"),
                            "Speed", StringValue
("ns3::ConstantRandomVariable[Constant=4.0]"),
                            "Bounds", StringValue
("0|200|0|200"));
mobility.Install (wifiStaNodes1);
```

After deploying the Wi-Fi AP-1 and its STAs, start configuring Wi-Fi PHY and MAC protocols on them using the following code snippet:.

1. First, create a common Wi-Fi channel between Wi-Fi AP-1 and its STAs:

```
YansWifiChannelHelper channel1 =
YansWifiChannelHelper::Default ();
YansWifiPhyHelper phy1;
phy1.SetChannel (channel1.Create ());
```

2. Next, configure the Wi-Fi standard, AP-specific Wi-Fi MAC by setting its SSID, and install the PHY and MAC protocols using the following code snippet. Collect Wi-Fi network devices for IP configuration:

```
    WifiHelper wifi1;
wifi1.SetStandard (WIFI_STANDARD_80211ac);
WifiMacHelper mac1;
Ssid ssid1 = Ssid ("Packt-1");
NetDeviceContainer apDevices1;
mac1.SetType ("ns3::ApWifiMac",
            "Ssid", SsidValue (ssid1));
apDevices1 = wifi1.Install (phy1, mac1, wifiApNode1);
```

3. Similarly, configure STA-specific Wi-Fi MAC by setting ActiveProbing for connecting with the AP SSID and install PHY and MAC protocols on all STAs using the following code snippets:

```
NetDeviceContainer staDevices1;
mac1.SetType ("ns3::StaWifiMac",
```

```
                        "Ssid", SsidValue (ssid1),
                        "ActiveProbing", BooleanValue (false));
        staDevices1 = wifi1.Install (phy1, mac1,
    wifiStaNodes1);
```

For now, we are done with setting up Wi-Fi LAN-1. Similarly, we set up Wi-Fi LAN-2 using the following code snippet:.

1.  After creating the necessary Wi-Fi APs and STAs for Wi-Fi LAN-2, start deploying Wi-Fi AP-2 at the location (300,300):

```
        NodeContainer wifiStaNodes2;
        wifiStaNodes2.Create (nWifi);
        NodeContainer wifiApNode2 = WiFiAPNodes.Get (1);
        MobilityHelper mobility1;
        Ptr<ListPositionAllocator> positionAlloc1 =
    CreateObject<ListPositionAllocator> ();
        positionAlloc1->Add (Vector (300, 300, 0.0));
        mobility1.SetPositionAllocator (positionAlloc1);
        mobility1.SetMobilityModel
    ("ns3::ConstantPositionMobilityModel");
        mobility1.Install (wifiApNode2);
```

2.  Then, deploy STAs randomly around Wi-Fi AP-2 at location (300,300) and configure a random walk mobility model for all STAs using the following code snippets:

```
        mobility1.SetPositionAllocator
    ("ns3::RandomDiscPositionAllocator",
                                    "X", StringValue ("300.0"),
                                    "Y", StringValue ("300.0"),
                                    "Rho", StringValue
    ("ns3::UniformRandomVariable[Min=0|Max=20]"));
        mobility1.SetMobilityModel
    ("ns3::RandomWalk2dMobilityModel",
                                    "Mode", StringValue
    ("Time"),
                                    "Time", StringValue
    ("100s"),
                                    "Speed", StringValue
    ("ns3::ConstantRandomVariable[Constant=4.0]"),
```

```
                                   "Bounds", StringValue
    ("200|400|200|400"));
      mobility1.Install (wifiStaNodes2);
```

3.  Next, similar to Wi-Fi LAN-1 AP and STAs, configure Wi-Fi LAN-2 AP and its STAs' standard, PHY, and MAC protocols and install them:

```
        WifiHelper wifi2;
      wifi2.SetStandard (WIFI_STANDARD_80211ac);
      YansWifiChannelHelper channel2
    =   YansWifiChannelHelper::Default ();
      YansWifiPhyHelper phy2;
      phy2.SetChannel (channel2.Create ());
      WifiMacHelper mac2;
      Ssid ssid2 = Ssid ("Packt-2");
      NetDeviceContainer staDevices2;
      mac2.SetType ("ns3::StaWifiMac",
                    "Ssid", SsidValue (ssid2),
                    "ActiveProbing", BooleanValue (false));
      staDevices2 = wifi2.Install (phy2, mac2,
    wifiStaNodes2);
        NetDeviceContainer apDevices2;
      mac2.SetType ("ns3::ApWifiMac",
                    "Ssid", SsidValue (ssid2));
      apDevices2 = wifi2.Install (phy2, mac2, wifiApNode2);
```

4.  After configuring and installing Wi-Fi-specific protocols, install `InternetStackHelper` on all nodes and configure IP addresses and routing rules based on their subnets using the following code snippets:

```
      InternetStackHelper stack;
      stack.Install (wifiApNode1);
      stack.Install (wifiApNode2);
      stack.Install (wifiStaNodes1);
      stack.Install (wifiStaNodes2);
      Ipv4AddressHelper address;
      address.SetBase ("10.1.1.0", "255.255.255.0");
      Ipv4InterfaceContainer p2pInterfaces;
      Ipv4InterfaceContainer staInterfaces1;
```

```
Ipv4InterfaceContainer staInterfaces2;
p2pInterfaces = address.Assign (p2pDevices);
address.SetBase ("10.1.3.0", "255.255.255.0");
address.Assign (apDevices1);
staInterfaces1=address.Assign (staDevices1);
address.SetBase ("10.1.4.0", "255.255.255.0");
address.Assign (apDevices2);
staInterfaces2=address.Assign (staDevices2);
Ipv4GlobalRoutingHelper::PopulateRoutingTables ();
```

For now, we are done with the successful deployment of Wi-Fi LAN-1 and LAN-2. Now, we will see how to test them using the following code snippet:.

1.  Start with testing whether all Wi-Fi STAs in a shop area are reachable from their respective AP using the ns-3 ping application. Here, each AP pings to all its STAs to confirm the connectivity:

```
for (int i=0;i<nWifi;i++)
{
  ApplicationContainer p;
  V4PingHelper ping = V4PingHelper (staInterfaces1.
GetAddress (i));
  ping.SetAttribute("Verbose",StringValue("true"));
  NodeContainer pingers;
  pingers.Add (wifiApNode1);
  p = ping.Install (pingers);
  p.Start (Seconds (2.0));
  p.Stop (Seconds (4.0));
}
for (int i=0;i<nWifi;i++)
{
  ApplicationContainer p;
  V4PingHelper ping = V4PingHelper (staInterfaces2.
GetAddress (i));
  ping.SetAttribute("Verbose",StringValue("true"));
  NodeContainer pingers;
  pingers.Add (wifiApNode2);
  p = ping.Install (pingers);
  p.Start (Seconds (2.0));
```

```
      p.Stop (Seconds (4.0));
   }
```

2. After testing the connectivity among APs and their STAs, test the connectivity between STAs in Wi-Fi LAN-1 and Wi-Fi LAN-2 using the ping application again. Start the ping after the previous test ends:

```
ApplicationContainer p;
V4PingHelper ping = V4PingHelper (staInterfaces2.
GetAddress (7));
ping.SetAttribute("Verbose",StringValue("true"));
NodeContainer pingers;
pingers.Add (wifiStaNodes1.Get(7));
p = ping.Install (pingers);
p.Start (Seconds (4.0));
p.Stop (Seconds (5.0));
```

3. Finally, test the Wi-Fi LAN-1 link speed between Wi-Fi STA-1 and STA-4 using the following code snippet:

```
uint8_t port;
Address localAddress (InetSocketAddress
(Ipv4Address::GetAny (), port));
PacketSinkHelper sink ("ns3::UdpSocketFactory",
localAddress);
ApplicationContainer apps_sink = sink.Install
(wifiStaNodes1.Get (4));
apps_sink.Start (Seconds (6.0));
apps_sink.Stop (Seconds (10.0));
OnOffHelper onoff ("ns3::UdpSocketFactory",
InetSocketAddress (staInterfaces1.GetAddress(4), port));
onoff.SetConstantRate (DataRate ("100Mbps"), 1000);
onoff.SetAttribute ("StartTime", TimeValue (Seconds
(6)));
onoff.SetAttribute ("StopTime", TimeValue (Seconds
(8)));
onoff.SetAttribute ("OnTime",  StringValue
("ns3::ConstantRandomVariable[Constant=1]"));
onoff.SetAttribute ("OffTime", StringValue
("ns3::ConstantRandomVariable[Constant=0]"));
```

```
ApplicationContainer apps_source = onoff.Install
(wifiStaNodes1.Get (0));
```

4.  After writing Wi-Fi LAN's testing code, let's close `main` by installing a flow monitor for performance evaluation, generating `wifi_animation.xml` to visualize simulation, and freeing all simulation resources using the following ns-3 code snippets:

```
Ptr<FlowMonitor> flowmon;
FlowMonitorHelper flowmonHelper;
flowmon = flowmonHelper.InstallAll ();
AnimationInterface anim ("wifi_animation.xml");
anim.EnablePacketMetadata ();
```

5.  Finally, stop your simulation and clean up simulation resources using the following code snippet:

```
Simulator::Stop (Seconds(15.0));
Simulator::Run ();
flowmon->SerializeToXmlFile ("wifilanspeed.xml", true,
true);
Simulator::Destroy ();
Simulator::Run ();
Simulator::Destroy ();
}
```

That's superb! We set up a Wi-Fi LAN simulation as per the hands-on activity successfully. Let's evaluate your Wi-Fi LAN simulation topology thoroughly in the next section.

## Wi-Fi LANs evaluation in ns-3

Let's run the `pkt_wifi_lans.cc` simulation at your ns-3 installation directory and observe the results using the following command:

**./ns3 run scratch/pkt_wifi_lans | tee wifiping**

Let's open the `wifiping` file to see the ping results. Since every STA is sending a ping to its AP, it contains ping results related to each STA IP address (*10.1.3.X* or *10.1.4.X*) and a message saying that *three packets were transmitted, three packets were received, and there was 0% packet loss*. That means we can confirm all STAs are reachable to their respective AP.

It is also possible to check specific ping results (shown in *Figure 4.6*) between AP-1 and its STAs using the following command, by passing the respective STA subnet IP address in the `grep` argument:

**cat wifiping |grep '10.1.3.5'**

It displays the following ping results (shown in *Figure 4.6*):

```
PING 10.1.3.5 - 56 bytes of data - 84 bytes including ICMP and IPv4 headers.
64 bytes from 10.1.3.5: icmp_seq=0 ttl=64 time=+17.7372ms
64 bytes from 10.1.3.5: icmp_seq=1 ttl=64 time=+1.83526ms
--- 10.1.3.5 ping statistics ---
```

Figure 4.6 – Ping results between AP-1 and its STAs

Similarly, ping results (shown in *Figure 4.7*) between AP-2 and its STAs can be checked using the STA's IP address:

```
cat wifiping |grep '10.1.4.5'
```

It displays the following ping results (shown in *Figure 4.7*):

```
PING 10.1.4.5 - 56 bytes of data - 84 bytes including ICMP and IPv4 headers.
64 bytes from 10.1.4.5: icmp_seq=0 ttl=64 time=+19.3242ms
64 bytes from 10.1.4.5: icmp_seq=1 ttl=64 time=+1.05251ms
--- 10.1.4.5 ping statistics ---
```

Figure 4.7 – Ping results between AP-2 and its STAs

Now, to test ping results (shown in *Figure 4.8*) between Wi-Fi LAN-1 (STA-7) and Wi-Fi LAN-2 (STA-7), use the following command with the STA-7 IP address:

```
cat wifiping |grep '10.1.4.9'
```

It displays the following ping results (shown in *Figure 4.8*):

```
PING 10.1.4.9 - 56 bytes of data - 84 bytes including ICMP and IPv4 headers.
64 bytes from 10.1.4.9: icmp_seq=0 ttl=62 time=+4.87117ms
--- 10.1.4.9 ping statistics ---
```

Figure 4.8 – Ping results between Wi-Fi LAN-1 and Wi-Fi LAN-2

As we configured ping for only 1 second, it is showing single ping response results.

We can also check the RTT results using the following command:

```
cat wifiping |grep 'rtt'
```

It displays the following RTT results (shown in *Figure 4.9*) during ping between AP and STAs:

```
rtt min/avg/max/mdev = 6/10.33/19/7.506 ms
rtt min/avg/max/mdev = 7/16/34/15.59 ms
rtt min/avg/max/mdev = 7/18.33/41/19.63 ms
rtt min/avg/max/mdev = 7/13/24/9.539 ms
```

Figure 4.9 – RTT results during ping between AP and STAs

Now, to check the Wi-Fi link speed between Wi-Fi STA-1 and STA-4, parse the flow monitor generated `wifilanspeed.xml` file:

```
python3 scratch/flowmon-parse-results.py wifilanspeed.xml
```

It displays the Wi-Fi LAN speed (shown in *Figure 4.10*) between its STAs:

```
FlowID: 1 (UDP 10.1.3.2/49153 --> 10.1.3.6/0)
       TX bitrate: 102804.11 kbit/s
       RX bitrate: 81210.37 kbit/s
       Mean Delay: 96.24 ms
       Packet Loss Ratio: 0.13 %
```

Figure 4.10 – Wi-Fi LAN link speed results

We can observe that AP is able to receive 82 Mbps and there is a packet loss of 0.13%. Similarly, we can configure the Wi-Fi LAN-1 link speed with the respective STA configuration.

> **Important note**
>
> Wi-Fi link speed mainly depends on standards (*802.11a/b/g/n/ac/ax*) and respective configurations. Here, we used the *802.11ac* standard for AP and STAs. If multiple stations are sharing the Wi-Fi link, the speed also will be shared among the STAs.

Finally, visualize the Wi-Fi LAN simulation using NetAnimator. Move to the `ns-allinone-3.36/netanim-3.108` folder, open `.NetAnim` using the following command, select `wifi-animation.xml`, and click **play** to visualize the Wi-Fi LAN simulation (shown in *Figure 4.11*):

```
./NetAnim
```

It displays Wi-Fi messages, UDP messages, and ARP messages exchanged during simulation (shown in *Figure 4.11*):

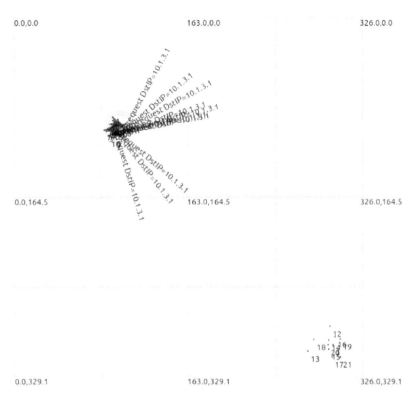

Figure 4.11 – Snapshot of Wi-Fi LAN simulation

While playing the simulation, we can visualize the following details:

- APs (node-0 and node-1) are placed at (100,100) and (300,300) locations
- STAs are moving around the AP
- APs periodically generate beacons
- The AP and STAs exchange association messages
- The AP and STAs exchange ping messages for 2 to 5 seconds
- Finally, STA-1 and STA-4 exchange UDP messages for 6 to 10 seconds

In this section, we discussed how to set up and evaluate our Wi-Fi LAN simulation using ns-3. We discussed how to test Wi-Fi link connectivity in a Wi-Fi LAN, how to evaluate Wi-Fi link speed, and the interworking of Wi-Fi LANs. In the next section, we are going to learn how to set up heterogeneous LAN simulations in ns-3.

# Step-by-step procedure for setting up heterogeneous LANs (CSMA and Wi-Fi LANs)

In many workplaces, schools, and organizations, we utilize multiple LANs. The main reasons for setting up multiple LANs include expanding network capacity and bringing network services closer to users. Usually, to expand any LAN connectivity capacity in quick time, network engineers can set up and install Wi-Fi APs. In this section, with a hands-on activity in ns-3, we show how to expand a CSMA LAN with two Wi-Fi networks (*Figure 4.12*):

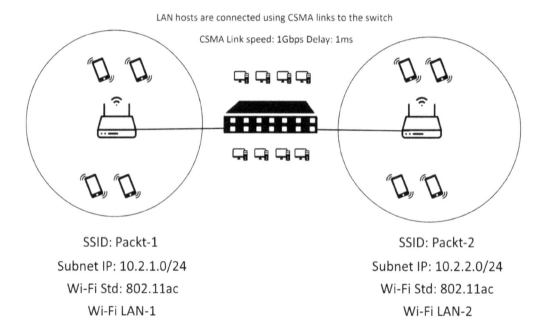

LAN hosts are connected using CSMA links to the switch

CSMA Link speed: 1Gbps Delay: 1ms

| SSID: Packt-1 | SSID: Packt-2 |
|---|---|
| Subnet IP: 10.2.1.0/24 | Subnet IP: 10.2.2.0/24 |
| Wi-Fi Std: 802.11ac | Wi-Fi Std: 802.11ac |
| Wi-Fi LAN-1 | Wi-Fi LAN-2 |

Figure 4.12 – Heterogeneous LANs (CSMA and Wi-Fi LANs)

The following are the major objectives of the hands-on activity:

- Set up a CSMA LAN with a link speed of 1Gbps and a delay of 1 ms.
- Extend the CSMA LAN with two Wi-Fi LANs.
- Configure the Wi-Fi standard as *802.11ac* and set up each Wi-Fi LAN with a unique SSID.
- Simulate mobility scenarios for Wi-Fi LANs. However, limit users' mobility within specific boundaries.
- Configure separate subnet IP addresses for each of the LANs.
- Test connectivity among all three LANs.

## Heterogeneous LAN simulation in ns-3

Let's set up a heterogeneous LAN simulation in a step-by-step procedure with the following ns-3 code snippets in pkt_het_lans.c:.

1.  Start with importing all the following necessary ns-3 packages:

    ```
    #include "ns3/core-module.h"
    #include "ns3/point-to-point-module.h"
    #include "ns3/network-module.h"
    #include "ns3/applications-module.h"
    #include "ns3/mobility-module.h"
    #include "ns3/csma-module.h"
    #include "ns3/internet-module.h"
    #include "ns3/yans-wifi-phy.h"
    #include "ns3/yans-wifi-helper.h"
    #include "ns3/wifi-net-device.h"
    #include "ns3/ssid.h"
    #include "ns3/netanim-module.h"
    using namespace ns3;
    ```

2.  In main(), start with creating the necessary number of ns-3 nodes for setting up the CSMA LAN using the following ns-3 code snippets. Here, you should note that from LAN-1's ns-3 nodes, two of the nodes (node-0 and node-23) will be used to configure the Wi-Fi APs later. That means that LAN-1 contains 22 hosts and 2 Wi-Fi APs:

    ```
    uint32_t nCsma = 24;
    uint32_t nWifi = 10;
    NodeContainer lan1;
    lan1.Create (nCsma);
    CsmaHelper csma;
    csma.SetChannelAttribute ("DataRate", StringValue
    ("100Mbps"));
    csma.SetChannelAttribute ("Delay", TimeValue
    (NanoSeconds (6560)));
    NetDeviceContainer lan1Devices;
    lan1Devices = csma.Install (lan1);
    ```

3.  After cabling the CSMA LAN, configure the LAN-1 host's IP address in a specific subnet (10.1.1.0) using the following ns-3 code snippets:

```
InternetStackHelper stack;
stack.Install (lan1);
address.SetBase ("10.1.1.0", "255.255.255.0");
Ipv4InterfaceContainer lan1Interfaces;
lan1Interfaces = address.Assign (lan1Devices);
```

So far, we set up the CSMA LAN and configure the IP addresses. Next, we begin setting up Wi-Fi LANs by extending existing the CSMA LAN at port-0 and port-23 with the installation of the respective Wi-Fi AP:

1.  Let's create the necessary NodeContainer and NetDeviceContainer for setting up Wi-Fi LANs:

```
NetDeviceContainer apDevices[2];
NetDeviceContainer staDevices[2];
NodeContainer wifiStaNodes[2];
Ipv4InterfaceContainer staInterfaces[2];
Ipv4InterfaceContainer apInterfaces[2];
NodeContainer wifiApNode[2];
```

2.  We set up two Wi-Fi LANs in the following loop with two iterations using the following ns-3 code snippets. In the following code, we deploy AP-1 at location (100,100) and AP-2 at location (300,300). Then, we deploy Wi-Fi LAN-1 STAs randomly around the AP-1, and Wi-Fi LAN-2 STAs around the AP-2. All Wi-Fi STAs are configured with a random walk mobility model with maximum speed allowed and specific boundaries around their AP's region:

```
for (int i=0;i<2;i++)
{
    wifiStaNodes[i].Create (nWifi);
    if (i==0)
    wifiApNode[i] = lan1.Get (0);
    if (i==1)
    wifiApNode[i] = lan1.Get (23);
    MobilityHelper mobility;
    Ptr<ListPositionAllocator> positionAlloc =
CreateObject<ListPositionAllocator> ();
    positionAlloc->Add (Vector (distance, distance,
0.0));
```

```
    mobility.SetPositionAllocator (positionAlloc);
    mobility.SetMobilityModel
("ns3::ConstantPositionMobilityModel");
    mobility.Install (wifiApNode[i]);
    if (i==0)
    {
      mobility.SetPositionAllocator
("ns3::RandomDiscPositionAllocator",
                              "X", StringValue
("100.0"),
                              "Y", StringValue
("100.0"),
                              "Rho", StringValue
("ns3::UniformRandomVariable[Min=0|Max=20]"));
      mobility.SetMobilityModel
("ns3::RandomWalk2dMobilityModel",
                              "Mode", StringValue
("Time"),
                              "Time", StringValue
("100s"),
                              "Speed", StringValue
("ns3::ConstantRandomVariable[Constant=2.0]"),
                              "Bounds", StringValue
("0|200|0|200"));
    }
```

3. So far, we have deployed AP-1 and its STAs. Next, we deploy AP-2 and its STAs using the following lines of code:

```
    else
    {
      mobility.SetPositionAllocator
("ns3::RandomDiscPositionAllocator",
                              "X", StringValue
("300.0"),
                              "Y", StringValue
("300.0"),
                              "Rho", StringValue
("ns3::UniformRandomVariable[Min=0|Max=20]"));
```

```
        mobility.SetMobilityModel
  ("ns3::RandomWalk2dMobilityModel",
                                "Mode", StringValue
  ("Time"),
                                "Time", StringValue
  ("100s"),
                                "Speed", StringValue
  ("ns3::ConstantRandomVariable[Constant=2.0]"),
                                "Bounds", StringValue
  ("200|400|200|400"));
    }
```

4.  Hey, be careful! We are still in the loop. Next, we configure and install Wi-Fi PHY and MAC protocols on APs and STAs by setting the respective SSIDs and STA's connection association mechanism:

```
    YansWifiChannelHelper channel =
  YansWifiChannelHelper::Default ();
    YansWifiPhyHelper phy;
    phy.SetChannel (channel.Create ());
    WifiMacHelper mac;
    Ssid ssid = Ssid ("Packt"+i);
    WifiHelper wifi;
    wifi.SetStandard (WIFI_STANDARD_80211ac);
    mac.SetType ("ns3::ApWifiMac",
                "Ssid", SsidValue (ssid));
    apDevices[i] = wifi.Install (phy, mac,
  wifiApNode[i]);
    mac.SetType ("ns3::StaWifiMac",
                "Ssid", SsidValue (ssid),
                "ActiveProbing", BooleanValue (false));
    staDevices[i] = wifi.Install (phy, mac,
  wifiStaNodes[i]);
```

5.  Now, we close the loop after configuring Wi-Fi LAN-1 and LAN-2 STAs with specific subnet IP addresses:

```
    InternetStackHelper stack;
    stack.Install (wifiStaNodes[i]);
    Ipv4AddressHelper address;
    std::ostringstream subnet;
```

```
    subnet<<"10.2."<<i+1<<".0";
    address.SetBase (subnet.str ().c_str (),
"255.255.255.0");
    address.Assign (apDevices[i]);
    staInterfaces[i]=address.Assign (staDevices[i]);
    distance=distance+200.0;
  }
Ipv4GlobalRoutingHelper::PopulateRoutingTables ();
```

That's all for setting up and configuring CSMA and Wi-Fi LANs successfully. Next, we test connectivity between the nodes in various LANs of our heterogeneous LAN setup:

1.  Install a UDP echo server application in Wi-Fi LAN-2 at STA-0 and configure its start and stop timings:

    ```
    UdpEchoServerHelper echoServer (9);
    ApplicationContainer serverApps = echoServer.Install
    (wifiStaNodes[1].Get (0));
    serverApps.Start (Seconds (1.0));
    serverApps.Stop (Seconds (10.0));
    ```

2.  Next, create and configure a UDP echo client application with the UDP echo server IP and port parameters:

    ```
    UdpEchoClientHelper echoClient (staInterfaces[1].
    GetAddress (0), 9);
    echoClient.SetAttribute ("MaxPackets", UintegerValue
    (1));
    echoClient.SetAttribute ("Interval", TimeValue (Seconds
    (1.0)));
    echoClient.SetAttribute ("PacketSize", UintegerValue
    (1024));
    ```

Let's test connectivity between the LANs by doing the following.

1. Check the connectivity between the CSMA LAN and Wi-Fi LAN-2 by testing whether Wi-Fi LAN-2's STA-0 is reachable and replying to the UDP echo client application installed on the CSMA LAN-1's host-5:

```
ApplicationContainer clientApps1 =
   echoClient.Install (lan1.Get (5));
clientApps1.Start (Seconds (2.0));
clientApps1.Stop (Seconds (3.0));
```

2. Check the connectivity between Wi-Fi LAN-1 and LAN-2 by testing whether Wi-Fi LAN-2's STA-0 is reachable and replying to the UDP echo client application installed on Wi-Fi LAN-1's host-5:

```
ApplicationContainer clientApps2 =
   echoClient.Install (wifiStaNodes[0].Get (5));
clientApps2.Start (Seconds (4.0));
clientApps2.Stop (Seconds (5.0));
```

3. After writing the heterogeneous LAN testing code, let's close main by installing a flow monitor for performance evaluation and freeing all simulation resources using the following ns-3 code snippets:

```
Simulator::Stop (Seconds(15.0));
Simulator::Run ();
flowmon->SerializeToXmlFile ("wifilanspeed.xml", true,
true);
Simulator::Destroy ();
Simulator::Run ();
Simulator::Destroy ();
}
```

That's superb. We have set up a heterogeneous LAN simulation with a hands-on activity using ns-3. Let's evaluate the heterogeneous LAN simulation topology thoroughly in the next section.

## Heterogeneous LAN evaluation in ns-3

Let's run the pkt_het_lans.cc simulation at your ns-3 installation directory and observe the results using the following command:

```
./ns3 run scratch/pkt_het_lans
```

To test the connectivity among all LANs, we can observe the following.

Let's check the connectivity between the CSMA LAN and Wi-Fi LAN-2 using the following flow monitor results (shown in *Figure 4.13*). We can observe that there is a successful UDP echo packets exchange between the Wi-Fi LAN-2's STA-0 (10.2.2.2) and CSMA LAN-1's host-5 (10.1.1.6):

```
At time +2s client sent 1024 bytes to 10.2.2.2 port 9
At time +2.01475s server received 1024 bytes from 10.1.1.6 port 49153
At time +2.01475s server sent 1024 bytes to 10.1.1.6 port 49153
At time +2.01987s client received 1024 bytes from 10.2.2.2 port 9
```

Figure 4.13 – Connectivity results between CSMA LAN-1 and Wi-Fi LAN-2

Similarly, check the connectivity between Wi-Fi LAN-1 and Wi-Fi LAN-2 using the following flow monitor results (shown in *Figure 4.14*). We can observe that there is a successful UDP echo packets exchange between Wi-Fi LAN-2's STA-0 (10.2.2.2) and Wi-Fi LAN-2's STA-5 (10.2.1.7):

```
At time +4s client sent 1024 bytes to 10.2.2.2 port 9
At time +4.01545s server received 1024 bytes from 10.2.1.7 port 49153
At time +4.01545s server sent 1024 bytes to 10.2.1.7 port 49153
At time +4.0204s client received 1024 bytes from 10.2.2.2 port 9
```

Figure 4.14 – Connectivity results between CSMA LAN-1 and Wi-Fi LAN-2

In this section, we set up and thoroughly evaluated a heterogeneous LAN simulation using ns-3. Specifically, we discussed how to connect Wi-Fi LANs and CSMA LANs, and how to test the connectivity between LANs and their link speeds. In the next section, we are going to learn how to simulate and evaluate a typical internet topology using ns-3.

# Step-by-step procedure for setting up and evaluating an internet simulation topology

We have successfully completed various hands-on activities related to setting up wired, Wi-Fi, and heterogeneous LANs. Now, we will see how to set up a typical internet topology, which comprises various access networks such as LANs, Wi-Fi LANs, and **Wide Area Networks** (**WANs**) connected through multiple routers, as seen in *Figure 4.15*:

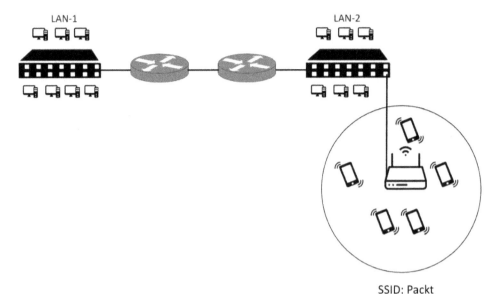

SSID: Packt
Subnet IP: 10.0.1.0/24
Wi-Fi Std: 802.11ac

Figure 4.15 – A typical internet setup in ns-3

In order to set up the internet in ns-3, the following hands-on activity is discussed in step-by-step procedures:

- Set up two LANs: LAN-1 and LAN-2. Assume these LANs are in two different countries.
- LAN-1 and LAN-2 hosts are connected using CSMA links to the switch.
- CSMA link speed of 1 Gbps and a delay of 1 ns.
- LANs are connected to the internet using routers through WAN links.
- WAN links are a P2P connection speed of 1 Gbps and a delay of 1 ns.
- LAN-2 is also interconnected with a Wi-Fi LAN (standard *802.11ac*).
- Connect and configure suitable link speed between LANs and WANs.
- Each LAN's host is configured in unique subnets.
- Connect these three LANs using WAN links.
- Simulate web browsing on the internet.
- Test connectivity between LAN-1 and LAN-2.

## Internet simulation in ns-3

Let's set up an internet simulation in a step-by-step procedure with the following ns-3 code snippets in `Internet.c`:

1.  First, import all the following necessary ns-3 packages:

    ```
    #include "ns3/core-module.h"
    #include "ns3/point-to-point-module.h"
    #include "ns3/network-module.h"
    #include "ns3/applications-module.h"
    #include "ns3/mobility-module.h"
    #include "ns3/csma-module.h"
    #include "ns3/internet-module.h"
    #include "ns3/yans-wifi-phy.h"
    #include "ns3/yans-wifi-helper.h"
    #include "ns3/wifi-net-device.h"
    #include "ns3/ssid.h"
    #include "ns3/flow-monitor-module.h"
    #include "ns3/netanim-module.h"
    using namespace ns3;
    ```

2.  In `main()`, start with creating the necessary number of ns-3 nodes for setting up LANs and routers. In LAN-2, ns-3 node-23 is reserved for Wi-Fi AP setup:

    ```
    uint32_t nCsma = 24;
    uint32_t nWifi = 10;
    NodeContainer lan1;
    lan1.Create (nCsma);
    NodeContainer lan2;
    lan2.Create (nCsma);
    NodeContainer routers;
    routers.Create(2);
    NodeContainer l1r = NodeContainer (lan1.Get (0),
    routers.Get(0));
    NodeContainer l2r = NodeContainer (lan2.Get (22),
    routers.Get(1));
    ```

3.  Next, set up LAN-1 and LAN-2 using CSMA channels of link speed 1 Gbps and delay 1 ns. Collect network devices for IP address configuration later:

```
CsmaHelper lancsma1;
lancsma1.SetChannelAttribute ("DataRate", StringValue
("1Gbps"));
lancsma1.SetChannelAttribute ("Delay", TimeValue
(NanoSeconds (1)));
CsmaHelper lancsma2;
lancsma2.SetChannelAttribute ("DataRate", StringValue
("1Gbps"));
lancsma2.SetChannelAttribute ("Delay", TimeValue
(NanoSeconds (1)));
NetDeviceContainer lan1Devices;
NetDeviceContainer lan2Devices;
lan1Devices = lancsma1.Install (lan1);
lan2Devices = lancsma2.Install (lan2);
```

4.  Next, establish WAN links between LAN-1 and router-1, LAN-2 and router-2, and between router-1 and router-2, using the following ns-3 code snippet. Collect the respective network devices for IP address configuration later:

```
PointToPointHelper wanlink1;
wanlink1.SetDeviceAttribute ("DataRate", StringValue
("1Gbps"));
wanlink1.SetChannelAttribute ("Delay", StringValue
("1ns"));
NetDeviceContainer wan1devs = wanlink1.Install (l1r);
PointToPointHelper wanlink2;
wanlink2.SetDeviceAttribute ("DataRate", StringValue
("1Gbps"));
wanlink2.SetChannelAttribute ("Delay", StringValue
("1ns"));
NetDeviceContainer wan2devs = wanlink2.Install (l2r);
PointToPointHelper wanlink3;
wanlink3.SetDeviceAttribute ("DataRate", StringValue
("1Gbps"));
wanlink3.SetChannelAttribute ("Delay", StringValue
("1ns"));
NetDeviceContainer wan3devs = wanlink3.Install
(routers);
```

5.  After connecting LANs with routers using WAN links, configure LAN-1 in subnet IP address (192.168.1.0), WAN link-1 in subnet IP address 192.168.2.0, WAN link-2 in subnet IP address 192.168.3.0, WAN link-3 in subnet IP address 192.168.4.0, and LAN-2 in subnet IP address 192.168.5.0, using the following ns-3 code snippets:

```
Ipv4AddressHelper address1;
address1.SetBase ("192.168.1.0", "255.255.255.0");
Ipv4InterfaceContainer lan1Interfaces;
lan1Interfaces = address1.Assign (lan1Devices);
Ipv4AddressHelper address2;
address2.SetBase ("192.168.2.0", "255.255.255.0");
Ipv4InterfaceContainer wan1Interfaces;
wan1Interfaces = address2.Assign (wan1devs);
Ipv4AddressHelper address3;
address3.SetBase ("192.168.3.0", "255.255.255.0");
Ipv4InterfaceContainer wan2Interfaces;
wan2Interfaces = address3.Assign (wan2devs);
Ipv4AddressHelper address4;
address4.SetBase ("192.168.4.0", "255.255.255.0");
Ipv4InterfaceContainer wan3Interfaces;
wan3Interfaces = address4.Assign (wan3devs);
Ipv4AddressHelper address5;
address5.SetBase ("192.168.5.0", "255.255.255.0");
Ipv4InterfaceContainer lan2Interfaces;
lan2Interfaces = address5.Assign (lan2Devices);
```

6.  Let's extend LAN-2 with a Wi-Fi LAN setup using the following code snippet. First, deploy an AP (LAN-2 ns-3 node-23) at location (100,100), and place STAs randomly around the AP. Besides this, configure all STAs with a random walk mobility model with the maximum speed allowed and specific boundaries around the AP:

```
NodeContainer wifiStaNodes;
NodeContainer wifiApNode;
NetDeviceContainer apDevices;
NetDeviceContainer staDevices;
Ipv4InterfaceContainer staInterfaces;
Ipv4InterfaceContainer apInterfaces;
wifiStaNodes.Create (nWifi);
wifiApNode = lan2.Get (23);
```

```
MobilityHelper mobility;
Ptr<ListPositionAllocator> positionAlloc =
CreateObject<ListPositionAllocator> ();
positionAlloc->Add (Vector (100, 100, 0.0));
mobility.SetPositionAllocator (positionAlloc);
mobility.SetMobilityModel
("ns3::ConstantPositionMobilityModel");
mobility.Install (wifiApNode);
mobility.SetPositionAllocator
("ns3::RandomDiscPositionAllocator",
                              "X", StringValue ("100.0"),
                              "Y", StringValue ("100.0"),
                              "Rho", StringValue
("ns3::UniformRandomVariable[Min=0|Max=20]"));
mobility.SetMobilityModel
("ns3::RandomWalk2dMobilityModel",
                              "Mode", StringValue
("Time"),
                              "Time", StringValue
("100s"),
                              "Speed", StringValue
("ns3::ConstantRandomVariable[Constant=2.0]"),
                              "Bounds", StringValue
("0|200|0|200"));
```

7.  After deploying AP and STAs, configure their PHY and MAC protocols by setting the SSID
    and STAs association mechanism using the following code snippets:

```
YansWifiChannelHelper channel
=    YansWifiChannelHelper::Default ();
YansWifiPhyHelper phy;
phy.SetChannel (channel.Create ());
WifiMacHelper mac;
Ssid ssid = Ssid ("Packt");
WifiHelper wifi;
wifi.SetStandard (WIFI_STANDARD_80211ac);
mac.SetType ("ns3::ApWifiMac",
                "Ssid", SsidValue (ssid));
apDevices = wifi.Install (phy, mac, wifiApNode);
```

```
mac.SetType ("ns3::StaWifiMac",
              "Ssid", SsidValue (ssid),
              "ActiveProbing", BooleanValue (false));
staDevices = wifi.Install (phy, mac, wifiStaNodes);
```

8.  Next, configure AP and its STAs with IP address in subnet 10.0.1.0:

```
Ipv4AddressHelper address;
address.SetBase ("10.0.1.0", "255.255.255.0");
address.Assign (apDevices);
staInterfaces=address.Assign (staDevices);
Ipv4GlobalRoutingHelper::PopulateRoutingTables ();
```

Well done. We have successfully completed setting up the internet comprising LANs and Wi-Fi with WAN links. Let's test our simulation setup for the connectivity of the internet host:

1.  First, we test web browsing between a Wi-Fi STA and a web server installed on LAN-1 using the following code snippets:

```
Ipv4Address serverAddress = lan1Interfaces.GetAddress
(1);
ThreeGppHttpServerHelper serverHelper (serverAddress);
ApplicationContainer serverApps1 = serverHelper.Install
(lan1.Get (1));
Ptr<ThreeGppHttpServer> httpServer = serverApps1.Get
(0)->GetObject<ThreeGppHttpServer> ();
PointerValue varPtr;
httpServer->GetAttribute ("Variables", varPtr);
Ptr<ThreeGppHttpVariables> httpVariables = varPtr.
Get<ThreeGppHttpVariables> ();
httpVariables->SetMainObjectSizeMean (1024000);
httpVariables->SetEmbeddedObjectSizeMean (102400);
ThreeGppHttpClientHelper clientHelper (serverAddress);
 ApplicationContainer clientApps1 = clientHelper.
Install (wifiStaNodes.Get (0));
clientApps1.Stop (Seconds (100));
```

2.  Next, we test a bidirectional messages exchange using the UDP echo client and server applications between LAN-1 and LAN-2 hosts (LAN-1's host-10 and LAN-2's host-5) connected using the internet:

```
UdpEchoServerHelper echoServer (9);
ApplicationContainer serverApps2 = echoServer.Install
(lan1.Get (10));
serverApps2.Start (Seconds (1.0));
serverApps2.Stop (Seconds (10.0));
UdpEchoClientHelper echoClient (lan1Interfaces.
GetAddress (10), 9);
echoClient.SetAttribute ("MaxPackets", UintegerValue
(100));
echoClient.SetAttribute ("Interval", TimeValue (Seconds
(0.01)));
echoClient.SetAttribute ("PacketSize", UintegerValue
(1024));
ApplicationContainer clientApps2 =
  echoClient.Install (lan2.Get (5));
clientApps2.Start (Seconds (2.0));
clientApps2.Stop (Seconds (10.0));
```

3.  Finally, let's close main by installing a flow monitor for performance evaluation and freeing all simulation resources using the following ns-3 code snippets:

```
Simulator::Stop (Seconds(15.0));
Simulator::Run ();
flowmon->SerializeToXmlFile ("wifilanspeed.xml", true,
true);
Simulator::Destroy ();
Simulator::Run ();
Simulator::Destroy ();
}
```

That's excellent! We successfully set up an internet simulation using ns-3. Let's evaluate our internet simulation topology thoroughly in the next section.

## Internet evaluation in ns-3

Let's run the `Internet.cc` simulation at your ns-3 installation directory and observe the results using the following command:

```
./ns3 run scratch/Internet
```

Let's check the connectivity between a web server installed on LAN-1's host-1 (`192.168.1.2`), and a web client installed on Wi-Fi LAN's STA-0 (`10.0.1.2`) using the following flow monitor results (shown in *Figure 4.16*):

```
FlowID: 1 (TCP 10.0.1.2/49153 --> 192.168.1.2/80)
        TX bitrate: 2.11 kbit/s
        RX bitrate: 57.29 kbit/s
        Mean Delay: 1.73 ms
        Packet Loss Ratio: 2.22 %
```
(a)

```
FlowID: 4 (TCP 192.168.1.2/80 --> 10.0.1.2/49153)
        TX bitrate: 3423.99 kbit/s
        RX bitrate: 3437.61 kbit/s
        Mean Delay: 200.39 ms
        Packet Loss Ratio: 0.00 %
```
(b)

Figure 4.16 – (a) TCP ACK exchange results during web browsing, (b) web objects download speed

---

**Important note**

In order to change browsing traffic, in simulation, change the HTTP main object mean and embedded object mean sizes.

---

By enabling the `ThreeGppHttpServer` and `ThreeGppHttpClient` log components, you can also view various HTTP messages exchanged between the web server and web client in terms of the generation of a variety of web objects and their sending and receiving details.

Finally, check the connectivity between LAN-1 and LAN-2 using the following flow monitor results (shown in *Figure 4.17*). We can observe that there is a successful UDP echo packets exchange between LAN-1's host-10 (`192.168.1.11`) and LAN-2's host-5 (`192.168.5.6`):

```
FlowID: 2 (UDP 192.168.5.6/49153 --> 192.168.1.11/9)
        TX bitrate: 935.11 kbit/s
        RX bitrate: 1093.41 kbit/s
        Mean Delay: 2.66 ms
        Packet Loss Ratio: 0.00 %
FlowID: 3 (UDP 192.168.1.11/9 --> 192.168.5.6/49153)
        TX bitrate: 1093.41 kbit/s
        RX bitrate: 1220.95 kbit/s
        Mean Delay: 2.78 ms
        Packet Loss Ratio: 0.00 %
```

Figure 4.17 – LAN-1 and LAN-2 connectivity test results

That's all! We have completed a detailed hands-on task related to setting up and evaluating the internet using ns-33's support for a variety of internet applications. Mainly, this hands-on activity helps you to get trained in how to set up and evaluate complex large network topologies in ns-3 systematically.

## Summary

In this chapter, we discussed various hands-on activities related to setting up LANs, Wi-Fi LANs, heterogeneous LANs, and the internet. Specifically, we discussed these hands-on activities related to real-time scenarios, which involved setting up networks in labs, workspaces, and hot spots, as well as how to extend existing networks. In a wireless network topology setup, you were introduced to using the ns-3 placement and mobility model for real-time wireless network scenarios. We also demonstrated important details to be visualized during wireless network simulations using NetAnimator.

These activities help you in setting up a variety of network topologies and evaluating them in terms of network connectivity, link speed, and testing a variety of ns-3 network applications over the simulation networks.

After practicing various interesting LANs simulation hands-on activities, in the next chapter, let's start exploring basic Wi-Fi technologies and wireless ad hoc network simulation setup and evaluation using ns-3.

# 5

# Exploring Basic Wi-Fi Technologies and Setting Up and Evaluating Wireless Ad Hoc Networks

In *Chapter 3* and *Chapter 4*, readers were already introduced to how to set up basic ns-3 Wi-Fi nodes, such as Wi-Fi STAs and APs, as well as how to set up Wi-Fi LANs. Specifically, in *Chapter 4*, users practiced in detail how to set up and simulate real-time LANs and internet networks using CSMA, P2P, and Wi-Fi technologies. However, some important details are missing related to wireless network simulations. In this chapter, we will fill this gap by discussing important parameters that influence wireless network simulations, such as the operating frequency or spectrum, signal losses, and interference issues in real-time wireless network operating environments. We will discuss how to simulate real-time Wi-Fi LANs and ad hoc network environments with the help of ns-3's supporting features, such as operating modes and frequencies, propagation models, and error-handling models. This chapter will also discuss various ns-3 supporting placement and mobility models for simulating ad hoc Wi-Fi networks and Wi-Fi LANs.

The first section of this chapter mainly focuses on aspects of the ns-3 Wi-Fi **Physical Layer** (**PHY**). It starts with discussing how to configure ns-3 node Wi-Fi standards, such as 802.11 a, b, and g; operating frequency bands (2.4 GHz/5 GHz) and channels; and operating modes (infrastructural or ad hoc). Then, it discusses how to simulate propagation loss and error handling in wireless network operating environments using ns-3 propagation models and error-handling models. The next section mainly focuses on the aspects of the ns-3 Wi-Fi **Medium Access Control** (**MAC**) layer. It starts with discussing how to configure and use important ns-3 features of Wi-Fi basic standards (802.11 a, b, and g) such as configuring the operating mode (infrastructural or ad hoc), channel reservation before frame transmissions, handling the transmission of large frames, using various rate control algorithms, and enabling **Quality-of-Service** (**QoS**) support. Then, in the third section, we discuss various ns-3 supporting placement and mobility models for simulating real-time wireless network topologies (infrastructural or ad hoc). Since ns-3 placement and mobility models are common to Wi-Fi or cellular network simulations, this will help users set up any mobile network simulation. In the past chapters, we discussed infrastructure-based Wi-Fi networks – Wi-Fi LANs. Now, we discuss how to set up wireless or **Mobile Ad Hoc Networks** (**MANETs**) using step-by-step procedures in the fourth section of this chapter. Finally, the last section discusses how to evaluate the performance of wireless ad hoc networks using suitable network applications and a variety of ad hoc network routing algorithms. In this chapter, we're going to cover the following main topics:

- Getting started with Wi-Fi nodes and channels in ns-3

- Exploring Wi-Fi – operating modes, rate control algorithms, and QoS support in ns-3

- Understanding placement models and mobility models

- Setting up wireless ad hoc networks

- Configuring and installing a variety of ad hoc network routing algorithms in ns-3

## Getting started with Wi-Fi nodes and channels in ns-3

The ns-3 Wi-Fi module provides support for both basic (802.11 a, b, or g) and advanced (802.11 n, ac, or ax) Wi-Fi standard simulations. ns-3 Wi-Fi module design broadly captures the following important aspects to accurately configure real-time Wi-Fi network simulations:

- Wi-Fi basic PHY modeling

- Wi-Fi advanced PHY (802.11 n, ac, or ax) feature modeling

- Wi-Fi low-level MAC layer function modeling

- Wi-Fi high-level MAC layer function modeling

In this section, we will start by discussing Wi-Fi basic PHY modeling and how to use it to set up Wi-Fi simulations more realistically. The ns-3 `WifiPhy` class models all basic PHY layer operations and functionalities. Basically, ns-3 PHY objects model the following important actions:

- In an ns-3 Wi-Fi node (STA or AP), the PHY object receives packets from the MAC layer and sends packets over the node's attached Wi-Fi channel (e.g., ns-3's `YansWifiChannel`).

- When receiving packets from a Wi-Fi channel attached to an ns-3 Wi-Fi node (STA or AP), the PHY object checks for packet errors and decides whether to send them for MAC layer processing. When receiving packets on a Wi-Fi channel, in order to model packet errors realistically, the ns-3 PHY layer can be configured to consider inputs from ns-3 propagation loss models, error models, and interference computing helpers.

- PHY layer operation states are modeled using ns-3's `WifiPhyStateHelper`. At any time, an ns-3 Wi-Fi node PHY layer can be in any of the following seven operation states:

  A.  `TX`: An ns-3 Wi-Fi node will be in the `TX` state after receiving a packet from the MAC layer until it finishes transmitting the signal on its attached channel.

  B.  `RX`: An ns-3 Wi-Fi node will be in the `RX` state when receiving a signal from its channel before handing it over to MAC layer processing.

  C.  `CCA_BUSY`: An ns-3 Wi-Fi node will be in the **Clear Channel Assessment** (**CCA**) state when checking the channel status (`BUSY` or `IDLE`).

  D.  `IDLE`: An ns-3 Wi-Fi node will be in an `IDLE` state when it is not transmitting, receiving, or making a CCA.

  E.  `SWITCHING`: When an ns-3 Wi-Fi node is switching its operating channel.

  F.  `SLEEP`: When an ns-3 Wi-Fi node is in power-saving mode (it does not transmit (`TX`) or receive (`RX`) any frames).

  G.  `OFF`: An ns-3 Wi-Fi node is off.

  In an ns-3 Wi-Fi node, the PHY layer operating state can be monitored. By default, its MAC layer listens to its underlying PHY layer state changes to initiate any **MAC Protocol Data Unit** (**MPDU**) transmission. It helps the MAC layer to avoid frame collisions.

Let's see how to configure an ns-3 Wi-Fi simulation more realistically with the help of ns-3 supporting Wi-Fi channels, propagation loss and delay models, and error models.

## ns-3 Wi-Fi PHY configuration

In past chapters, we have only used basic Wi-Fi PHY configurations to quickly set up a Wi-Fi network. To configure an ns-3 Wi-Fi network simulation more realistically, we need to use the following important PHY layer configuration on ns-3 Wi-Fi nodes:

- Include the following necessary ns-3 Wi-Fi packages:

```
#include "ns3/yans-wifi-phy.h"
#include "ns3/yans-wifi-helper.h"
#include "ns3/wifi-net-device.h"
#include "ns3/ssid.h"
```

- Create the required Wi-Fi APs and STAs using ns-3's `NodeContainers` as follows:

```
uint32_t nWifi = 10;
NodeContainer WiFiAPNodes;
WiFiAPNodes.Create (1);
```

```
NodeContainer wifiStaNodes;
wifiStaNodes.Create (nWifi);
```

- Set the Wi-Fi network standard, such as 802.11 a, b, g, n, ac, or ax, using the following lines of code:

```
WifiHelper wifi;
wifi.SetStandard (WIFI_STANDARD_80211a);
```

- Configure a specific Wi-Fi operating channel between an AP and its STAs using the following code:

```
YansWifiPhyHelper phyHelper;
phyHelper.Set ("ChannelSettings", StringValue ("{36, 20,
BAND_5GHZ, 0}"));
```

This sets the operating channel to 36, the channel width to 20 MHz, and the operating frequency band to 5 GHz. In the case of multiple APs, it is necessary to set neighbor APs with unique orthogonal (non-interfering) channels.

Before now, we have configured the Wi-Fi node standard, operating frequency band, channel width, and channels. To simulate real-time Wi-Fi network environments, it is also necessary to configure channel propagation loss and delay models and error models for ns-3 Wi-Fi nodes. Next, we will introduce important ns-3 supporting propagation loss, delay, and error models and how to configure them for our Wi-Fi network simulations.

## ns-3 supporting propagation loss, delay, and error rate models

Unlike wired network transmissions, wireless transmissions severely suffer from signal propagation losses due to a variety of obstacles between senders and receivers and interference from neighbor wireless radio transmissions. It leads to errors in packet reception at Wi-Fi nodes and variations within the dynamic transmission speed. The PHY layer of ns-3 Wi-Fi nodes, by default, computes the interference faced during transmission using ns-3's InterferenceHelper class. An InterferenceHelper object checks all incoming packets over a channel and calculates the probability of error values in received packets, and passes these values to WifiPhy. Equally, for any wireless network simulations, it is also necessary to consider propagation loss issues and resulting errors explicitly by configuring suitable propagation loss and error models. In ns-3, without much effort, these issues can be simulated by configuring the PHY layer channel of ns-3 Wi-Fi nodes with propagation loss and error models.

### ns-3 propagation delay and loss models

In ns-3, it is possible to simulate wireless simulations that match real-time wireless networks by using suitable propagation delay and loss models.

## ns-3 PropagationDelay models

Wi-Fi signal loss depends on the speed at which the signal travels and the distance between nodes. ns-3 supports the following signal delay models to simulate these losses:

- `ConstantSpeedPropagationDelayModel`
- `RandomPropagationDelayModel`

Users can simulate propagation delay losses by using the following lines of code:

```
YansWifiChannelHelper wifiChannel;
wifiChannel.SetPropagationDelay
("ns3::ConstantSpeedPropagationDelayModel");
```

## ns-3 PropagationLoss models

Besides propagation delay modeling, ns-3 supports a rich set of propagation loss models, which compute the signal receiving power by considering inputs such as the signal transmission power, the distance between the sender and the receiver, and a variety of potential obstacles between the sender and the receiver, such as buildings, rooms, walls, and wall types.

Here are a few of ns-3's `PropagationLoss` models: `FixedRss`, `FriisPropagation`, `JakesPropagation`, `LogDistancePropagation`, `NakagamiPropagation`, `RandomPropagation`, `RangePropagation`, `HybridBuildingsPropagation`, and `OhBuildings`.

For example, `LogDistancePropagation` can be configured in simulations to consider propagation loss due to a wide range of environmental conditions in which there is no clear path between the transmitter and the receiver Wi-Fi nodes.

In a wireless network environment simulation, users can configure multiple ns-3 propagation loss models as a chain. In the case of configuring multiple propagation loss models, the receiving power computation considers all the model inputs in a chain. This helps users to simulate a variety of signal-fading effects. Let's see how easy it is to simulate propagation loss using the following line of code in ns-3.

For example, to consider the free space between the Wi-Fi sender and the receivers and simulate the propagation loss, in ns-3, this can be done with the following simple line of code:

```
wifiChannel.AddPropagationLoss
("ns3::FriisPropagationLossModel");
```

Discussing further details on implementing various propagation loss models is beyond the scope of this book, but we recommend readers explore ns-3 propagation loss models in the ns-3 model documentation.

### ns-3 error rate models

In order to incorporate error handling into Wi-Fi network simulations, ns-3 Wi-Fi nodes decide whether a packet is successfully received or not based on the **Signal to Interference & Noise Ratio** (**SINR**) and **Packet Error Rate** (**PER**). As a packet's SINR is computed with the help of ns-3's `InterferenceHelper` object, to model the calculation of the PER realistically, ns-3 wireless networks should be configured with error models. Hence, ns-3 Wi-Fi nodes take inputs from `ErrorRateModels` and decide on a suitable **Modulation Coding Scheme** (**MCS**), and the MCS indicates the number of bits carried over a symbol of the input signal. When the Wi-Fi experiences a higher amount of noise and a higher number of errors, it is good to transmit signals with a lower MCS. Otherwise, receivers cannot decode any packets and may drop the packets. Overall, to simulate the PER realistically for a packet (or signal), ns-3 Wi-Fi nodes consider inputs from `ErrorRateModels`, the SINR, and the MCS.

In ns-3, the following error rate models are supported:

- For 802.11 n, ac, and ax, `ns3::TableBasedErrorRateModel` is supported
- For 802.11 a or g, `ns3::YansErrorRateModel` is supported
- For 802.11b: `ns3::DsssErrorRateModel` is supported
- For 802.11a, g, or n OFDM modes, `ns3::NistErrorRateModel` is also supported

Let's see how to configure error rate models in ns-3 using the following lines of code:

```
YansWifiPhyHelper phyHelper;
phyHelper.SetErrorModel("ns3::NistErrorRateModel");
```

For details of how these error rate models work, we suggest users go through the ns-3 documentation.

In order to deploy real-time Wi-Fi simulations in ns-3, users are suggested to use all the features of ns-3 related to the Wi-Fi PHY layer. In the next section, we will discuss all the features of ns-3 related to the Wi-Fi MAC layer.

## Exploring Wi-Fi – operating modes, rate control algorithms, and QoS support using ns-3

In order to match Wi-Fi simulation operations to real-time Wi-Fi operations, the ns-3 Wi-Fi module implements MAC functionalities or operations on ns-3 Wi-Fi nodes (AP or STAs) in two levels: low-level MAC and high-level MAC. Broadly, low-level MAC handles time-critical functions or operations, such as channel access functions, frames protection, and ACK generation. On the other hand, high-level or upper MAC handles non-time-critical operations, such as Wi-Fi beacon generation, management frame generation, such as connection handling (association messages), control frame generation, such as **Request-To-Send/Clear-To-Send** (**RTS/CTS**) frames, and data rate control operations.

Note that ns-3 Wi-Fi MAC does not support the following operations:

- STA authentication

- Data encryption

- 802.11 **Point Coordination Function (PCF)** or **Hybrid Coordination Function Controlled Channel Access (HCCA)**

- **Channel Switch Announcement (CSA)**

- Power saving

- Handovers or roaming between APs

- Modeling process delays

In this section, we will mainly discuss the following ns-3 Wi-Fi MAC features and operations and how to use them in our simulations:

- Wi-Fi operating modes (infrastructural or ad hoc)

- Data transmission rate control between ns-3 Wi-Fi nodes

- Enabling QoS support in a Wi-Fi network

## Wi-Fi operating modes (infrastructural or ad hoc)

Usually, we set up Wi-Fi networks to provide wireless LAN facilities or handle emergency or rescue-handling situations. Setting up a wireless LAN involves planning a centralized controller, such as a Wi-Fi AP deployment, and connecting the Wi-Fi APs using CSMA or P2P channels to handle any Wi-Fi STAs traffic in an environment (as we discussed for a Wi-Fi LAN setup in *Chapter 3* and *Chapter 4*).

### Infrastructure mode

In infrastructure mode, there are two types of ns-3 Wi-Fi nodes: APs and STAs. APs are central entities that announce the Wi-Fi network visibility by sending their SSIDs in beacons, allowing STAs to connect to a Wi-Fi LAN using management frames, exchanging any traffic between the STAs connected to them, and allowing the STAs to connect to the other networks (other LANs or the internet) through Wi-Fi LANs.

Let's quickly revisit how to configure ns-3 Wi-Fi node MAC (AP and STAs) to operate in infrastructure mode using the following configurations.

Use the following MAC configuration for an AP setup:

```
WifiMacHelper mac;
Ssid ssid = Ssid ("pack-1");
```

```
mac.SetType ("ns3::ApWifiMac",
             "Ssid", SsidValue (ssid1));
```

Use the following MAC configuration for an STA setup in infrastructure mode:

```
mac.SetType ("ns3::StaWifiMac",
             "Ssid", SsidValue (ssid),
             "ActiveProbing", BooleanValue (false));
```

In infrastructure operating mode, ns-3 AP nodes can handle the following tasks:

- Generating beacons on a particular Wi-Fi channel

- Generating the necessary management and control frames

- Handling association request messages (connection-related) from Wi-Fi STAs using the configured Wi-Fi channel

- Following Wi-Fi MAC channel access rules

ns-3 Wi-Fi STAs can handle the following tasks:

- They listen to beacons and initiate connection requests through an AP-specific channel.

- For association with an AP, if the `ActiveProbing` mode is not enabled, by default, an STA will scan various channels in the `PassiveScanning` mode. In `PassiveScanning`, STAs will generate PROBE request messages for various APs and collect the respective PROBE response messages from various APs. Then, an STA will select a particular AP to connect with the AP by exchanging associated request and response messages.

- If there are connection failures, STAs retry the connection with the AP or other APs on their specific channels.

- They follow Wi-Fi MAC channel access rules.

- The current ns-3 implementation does not support roaming (handovers) between APs.

## Ad hoc mode

On other hand, to handle message handling without the given infrastructure, there will be no APs set up, but Wi-Fi STAs themselves will connect to each other to exchange traffic. In order to simulate these networks, we need to configure ns-3 Wi-Fi nodes in an ad hoc network (an **Independent Basic Service Set** (**IBSS**)), and their MAC layer should be configured as ns3::AdhocWifiMac. In ad hoc mode, Wi-Fi nodes do not perform any beacon generation, passive scanning, or active scanning operations. We can configure ns-3 Wi-Fi node MAC to operate in ad hoc mode as follows:

```
mac.SetType ("ns3::AdhocWifiMac");
```

## Channel access

Whether an ns-3 Wi-Fi node is in infrastructure or ad hoc mode, before any frame transmission, it follows the 802.11 **Distributed Coordination Function** (**DCF**), similar to real-time Wi-Fi nodes.

In ns-3, Wi-Fi nodes use the DCF to decide when a node can get channel access for any Wi-Fi frame transmission (management, control, or data). The ns-3 DCF implementation works as follows: an ns-3 Wi-Fi node transmits a frame when the medium is idle and remains idle for a **DCF Inter-Frame Spacing** (**DIFS**) or **Extended IFS** (**EIFS**), and the backoff counter reaches 0. Meanwhile, if the medium is busy again, the ns-3 Wi-Fi node enters a random backoff procedure, and then the DCF process repeats.

### RTS/CTS for hidden terminals

In wireless networks, all Wi-Fi nodes may not be visible to all other nodes. This leads to frame collisions at destination nodes due to simultaneous transmissions on a channel from hidden nodes to the destination nodes. To protect frames from collisions, 802.11 offers RTS/CTS control frame exchanges before actual data frame transmissions take place. ns-3 also supports RTS/CTS support for frame protection. This can be enabled in a Wi-Fi network simulation using the following lines of code in ns-3:

```
Config::SetDefault
("ns3::WifiRemoteStationManager::RtsCtsThreshold", StringValue
("1200"));
```

This means for any frame over the size of 1200 for an ns-3 Wi-Fi node, an RTS/CTS frame exchange occurs.

### Handling large frame transmissions

On the other hand, in wireless networks, due to noisy channels and errors, large frames have a higher probability of being corrupted than smaller frames. To address this issue, 802.11 supports frame fragmentation mode to fragment larger frames into smaller frames based on a fragmentation threshold. All of these smaller fragments are transmitted in a single DCF access. The ns-3 Wi-Fi module also provides this feature by using the following line of code:

```
Config::SetDefault
("ns3::WifiRemoteStationManager::FragmentationThreshold",
StringValue ("2000"));
```

This means for any frame over the size of 2000, for an ns-3 Wi-Fi node, the larger frame will be divided into smaller frames and they will be transmitted in a single DCF access.

## ns-3 support for rate control algorithms

In wireless networks, Wi-Fi nodes must adapt their data transmission rates dynamically to avoid frame errors, and reduce the PER. Hence, Wi-Fi nodes use rate control algorithms to decide the suitable data transmission rates, which consider the dynamic channel conditions (SNR variations), ACKs, and error rates. In the literature, there are many rate control algorithms proposed for providing optimal data transmission rates between Wi-Fi nodes. In ns-3, the following rate control algorithms are supported:

- `AarfWifiManager`
- `AarfcdWifiManager`
- `CaraWifiManager`
- `MinstrelWifiManager`
- `MinstrelHtWifiManage`
- `OnoeWifiManager`
- `AmrrWifiManager`
- `IdealWifiManager`
- `ArfWifiManager`
- `OnoeWifiManager`
- `ConstantRateWifiManager`

Let's see how to configure a rate control algorithm for Wi-Fi nodes in ns-3 using the following lines of code:

```
WifiHelper wifi;
wifi.SetRemoteStationManager ("ns3::ArfWifiManager");
```

## ns-3 QoS support

In real-time Wi-Fi networks, by enabling QoS support on Wi-Fi nodes, it is possible to classify traffic flows into the following four **access categories** (**ACs**): voice traffic (`AC_VO`), video traffic (`AC_VI`), best-effort traffic (`AC_BE`), and background traffic (`AC_BK`). Hence, it is possible to handle these ACs based on their priority rules. In ns-3 wireless network simulations, it is also possible to enable QoS for all types of Wi-Fi nodes, either in infrastructural or ad hoc networks.

For example, let's see how to enable QoS for Wi-Fi APs in ns-3 using the following code snippet:

```
WifiMacHelper mac;
mac.SetType ("ns3::ApWifiMac",
```

```
"Ssid", SsidValue (ssid),
"QosSupported", BooleanValue (true));
```

Similarly, for STAs, we can enable QoS using the following lines of code:

```
mac.SetType ("ns3::StaWifiMac",
"Ssid", SsidValue (ssid),
"QosSupported", BooleanValue (true),
"ActiveProbing", BooleanValue (false));
```

Before now, we have discussed how to enable QoS support on Wi-Fi nodes. Equally, when testing QoS applications, it is necessary to configure a QoS tag for ns-3 application traffic flows. While ns-3 Wi-Fi nodes handle **MAC Service Data Units (MSDUs)** from a higher layer (the IP), they look into the **Type of Service (ToS)** and set a suitable AC for the packets. Hence, in our simulations, we need to set the ToS for traffic flows.

For example, in a Wi-Fi network simulation, every STA connected to the AP needs to be configured with four different AC traffic flows. Let's see how to do this in our simulation, using the following code snippet for configuring the ToS and QoS tags for ns-3 applications:

```
ApplicationContainer sources, sinks;
std::vector<uint8_t> tos_acs = {0x70, 0x28, 0xb8, 0xc0};  //
AC_BE, AC_BK, AC_VI, AC_VO
uint32_t port = 9;
for (uint32_t i = 0; i < nWifi; ++i)
  {
    for (uint8_t tos_ac : tos_acs)
      {
        auto ipv4 = wifiApNode.Get (0)->GetObject<Ipv4> ();
        const auto address = ipv4->GetAddress (1, 0).GetLocal
();
        InetSocketAddress sinkSocket (address, port++);
        sinkSocket.SetTos (tosValue);
        OnOffHelper onOffHelper ("ns3::UdpSocketFactory",
sinkSocket);
        onOffHelper.SetAttribute ("OnTime", StringValue
("ns3::ConstantRandomVariable[Constant=1]"));
        onOffHelper.SetAttribute ("OffTime", StringValue
("ns3::ConstantRandomVariable[Constant=0]"));
        sources.Add (onOffHelper.Install (wifiStaNodes.Get
(index)));
```

```
        PacketSinkHelper packetSinkHelper
("ns3::UdpSocketFactory", sinkSocket);
        sinks.Add (packetSinkHelper.Install (wifiApNodes.Get
(0)));
    }
  }
```

As we discussed, ns-3 users can configure MAC features, such as rate control algorithms, QoS, RTS/ CTS, and fragmentation modes to simulate either Wi-Fi LANs or ad hoc networks that correspond to real-time Wi-Fi networks. In the next section, we will discuss how to simulate wireless network topologies using ns-3 placement and mobility modules.

# Understanding placement models and mobility models

In wireless network simulations, it is very important to set up network topologies that match a real-time environment setup in terms of the equipment deployment locations and user mobility patterns. In order to set up either Wi-Fi networks or cellular networks, ns-3 provides a rich set of placement and mobility models. In this section, we will discuss how to use and configure various ns-3 placement and mobility models in our network simulations.

Let's start by exploring the following ns-3 placement models:

- `ListPositionAllocator`
- `GridPositionAllocator`
- `RandomDiscPositionAllocator`
- `RandomRectanglePositionAllocator`
- `GridBuildingAllocator` and `RandomRoomPositionAllocator`

## ListPositionAllocator

In wireless network simulations, it is necessary to set specific locations for Wi-Fi APs or cellular base stations so that it is possible to deploy them according to real-time site survey locations. In ns-3, location-specific network equipment deployments can easily be made using `ListPositionAllocator`. For example, to deploy two Wi-Fi APs at locations of (100,100) and (500,500) with a constant AP mobility model, you can do so in ns-3 with the following lines of code:

```
uint32_t nWifi = 2;
NodeContainer WiFiAPNodes;
WiFiAPNodes.Create (2);
MobilityHelper mobility;
```

```
Ptr<ListPositionAllocator> positions =
CreateObject<ListPositionAllocator> ();
  positions->Add (Vector (100.0, 100.0, 0.0));
  positions->Add (Vector (500.0, 500.0, 0.0));
  mobility.SetPositionAllocator (positions);
  mobility.SetMobilityModel
("ns3::ConstantPositionMobilityModel");
  mobility.Install (wifiAPNodes);
```

## GridPositionAllocator

In some wireless network simulation use cases, such as for auditoriums and stadiums, it is possible to assume Wi-Fi nodes are deployed in a grid topology. This means that Wi-Fi nodes are deployed in multiple rows. In each row, the distance between Wi-Fi nodes is fixed, as is the distance between rows. For example, we want to deploy 16 Wi-Fi STAs (a distance between STAs of 5 m and a distance between rows of 5 m) as follows:

| STA-1 | STA-2 | STA-3 | STA-4 |
| STA-5 | STA-6 | STA-7 | STA-8 |
| STA-9 | STA-10 | STA-11 | STA-12 |
| STA-13 | STA-14 | STA-15 | STA-16 |

In ns-3, using GridPositionAllocation, this can be done with the following lines of code:

```
uint32_t nWifi = 16;
NodeContainer WiFiStaNodes;
WiFiStaNodes.Create (nWifi);
MobilityHelper mobility;
  mobility.SetPositionAllocator ("ns3::GridPositionAllocator",
"MinX", DoubleValue (0.0), "MinY", DoubleValue (0.0), "DeltaX",
DoubleValue (5.0), "DeltaY", DoubleValue (5.0),
  "GridWidth", UintegerValue (4),
  "LayoutType", StringValue ("RowFirst"));
  mobility.
SetMobilityModel  ("ns3::ConstantPositionMobilityModel");
  mobility.Install (wifiStaNodes);
```

Here, MinX and MinY are the first STA position, DeltaX sets the distance between STAs, and DeltaY sets the distance between rows, while GridWidth indicates the number of STAs per row.

## RandomDiscPositionAllocator

Before now, we have deployed ns-3 nodes at specific locations, but in wireless network simulations, it is very common that wireless nodes are located at random locations in an environment or place. In ns-3, deploying ns-3 nodes at random locations can be done using random position allocator models. Let's see how to simulate the following in a wireless network:

- An AP is deployed at the location of (100,100)
- STAs should be randomly deployed around the AP
- No STAs should be deployed at a radius greater than 30 from the AP location

The preceding scenario can be simulated using the following lines of code:

```
uint32_t nWifi = 10;
NodeContainer WiFiAPNodes;
NodeContainer WiFiStaNodes;
WiFiAPNodes.Create (1);
WiFiStaNodes.Create (nWifi);
MobilityHelper mobility;
Ptr<ListPositionAllocator> positions
=    CreateObject<ListPositionAllocator> ();
positions->Add (Vector (100.0, 100.0, 0.0));
mobility.SetMobilityModel
("ns3::ConstantPositionMobilityModel");
mobility.Install (wifiAPNodes);
mobility.SetPositionAllocator
("ns3::RandomDiscPositionAllocator", "X", StringValue
("100.0"), "Y", StringValue ("100.0"),
 "Rho", StringValue
("ns3::UniformRandomVariable[Min=0|Max=30]"));
mobility.SetMobilityModel
("ns3::ConstantPositionMobilityModel");
mobility.Install (wifiStaNodes);
```

Here, the Disc center location is defined by X and Y coordinates, and the maximum radius from the center at which STAs can be deployed is constrained by Rho.

## RandomRectanglePositionAllocator

Next, we will see how to deploy STAs randomly within a box area. Using ns-3's RandomRectanglePositionAllocator, let's see how to deploy STAs within a box area. The min X and min Y are (-100,-100) and the max X and max Y are (100,100) using the following code snippet:

```
mobility.SetPositionAllocator
("ns3::RandomRectanglePositionAllocator","X", StringValue
("ns3::UniformRandomVariable[Min=-100.0|Max=100.0]"),
"Y", StringValue ("ns3::UniformRandomVariable[Min=-
100.0|Max=100.0]"));
mobility.Install (wifiStaNodes);
```

Next, we will discuss another interesting and real-time placement model called the BuildingPosition mobility model.

## GridBuildingAllocator and RandomRoomPositionAllocator

ns-3 provides the GridBuildingAllocator and RandomRoomPositionAllocator models to simulate real-time Wi-Fi deployments in buildings such as offices, schools, and so on. For example, let's see how to create two buildings, with each building containing two floors, and each floor containing two rooms on the *x*-axis and four rooms on the *y*-axis:

```
Ptr<GridBuildingAllocator>  gridBuildingAllocator;
gridBuildingAllocator = CreateObject<GridBuildingAllocator>
();
gridBuildingAllocator->SetAttribute ("GridWidth",
UintegerValue (2));
gridBuildingAllocator->SetAttribute ("LengthX", DoubleValue
(50));
gridBuildingAllocator->SetAttribute ("LengthY", DoubleValue
(50));
gridBuildingAllocator->SetAttribute ("DeltaX", DoubleValue
(2));
gridBuildingAllocator->SetAttribute ("DeltaY", DoubleValue
(2));
gridBuildingAllocator->SetAttribute ("Height", DoubleValue
(20));
gridBuildingAllocator->SetBuildingAttribute ("NRoomsX",
UintegerValue (2));
```

```
    gridBuildingAllocator->SetBuildingAttribute ("NRoomsY",
UintegerValue (4));
    gridBuildingAllocator->SetBuildingAttribute ("NFloors",
UintegerValue (2));
    gridBuildingAllocator->SetAttribute ("MinX", DoubleValue
(0));
    gridBuildingAllocator->SetAttribute ("MinY", DoubleValue
(0));
    gridBuildingAllocator->Create (2);
```

Here, first, we defined each grid size (`GridWidth`) for installing a building. Then, we configured the grid length in terms of the *x*- and *y*-axes, and the space between grids. Later, we defined the number of rooms in the X and Y directions, and the number of floors of a building. Finally, we configured the number of buildings to be created with the preceding dimensions.

Now, let's see how to deploy ns-3 nodes in random rooms of these buildings using the following code snippet:

```
Ptr<RandomRoomPositionAllocator> position =
CreateObject<RandomRoomPositionAllocator> ();
```

Hey! We have explored various ns-3 placement models and learned how to use them in our simulations. As we observed with random placement models, ns-3 nodes will be placed at random locations, but users need not worry as these random locations can easily be monitored using the following lines of code:

```
for (NodeContainer::Iterator i = WifiStaNodes.Begin ();
     i != WifiStaNodes.End (); ++i)
  {
    Ptr<Node> object = *j;
    Ptr<MobilityModel> position = object-
>GetObject<MobilityModel> ();
    Vector pos = position->GetPosition ();
    std::cout << "x=" << pos.x << ", y=" << pos.y << ", z="
<< pos.z << std::endl;
  }
```

In all these placement models, ns-3 Wi-Fi nodes are static (so do not move) during the simulation. Next, we will discuss how to simulate various mobility patterns in ns-3.

# ns-3 mobility models

ns-3 provides the following mobility models to simulate a variety of mobility patterns in Wi-Fi or cellular networks:

- `RandomWalk2dMobilityModel`
- `RandomWaypointMobilityModel`
- `RandomWalk2dOutdoorMobilityModel`

## *RandomWalk2dMobilityModel*

This model is suitable for simulating users' mobility in shopping malls, airports, exhibitions, and so on. For example, we can simulate the following use case using `RandomWalk2dMobilityModel`:

- Mobile users hold mobile phones and randomly move around an area
- Mobile users do not cross the boundaries of an area of `(0,0)` and `(200,200)`
- Mobile users may spend a 60-second fixed duration in one place and then they may move to another place at a random speed

We can simulate the preceding use case using the following ns-3 code snippet:

```
Mobility mobility;
mobility.SetMobilityModel ("ns3::RandomWalk2dMobilityModel",
                           "Mode", StringValue ("Time"),
                           "Time", StringValue ("60s"),
                           "Speed", StringValue
("ns3::ConstantRandomVariable[Constant=2.0]"),
                           "Bounds", StringValue
("0|200|0|200"));
```

The Wi-Fi node mobility can be defined based on `Mode` `(Distance or Time)`. In the `Distance` mode, the user changes their mobility pattern (at a random speed and with a random direction) after moving a specific distance. In the case of the `Time` mode, the user changes their mobility pattern after moving for a specific duration. `Bounds` defines the users' mobility boundaries.

## *RandomWaypointMobilityModel*

This model helps us to simulate the following types of mobility scenarios:

- In a specific area, there are many locations. Various users may visit various locations.
- A user starts moving towards a location at a random speed, and after reaching the particular location, they wait for a certain duration (a pause).

- After this pause, the user can choose another location and start moving toward the new location at a random speed.

Unlike `RandomWalk2dMobilityModel`, this model does not have explicit boundaries, but users need to configure set point positions using `PostionAllocator`.

Using the following code snippet, we can simulate these mobility patterns:

```
MobilityHelper waypoint;
ObjectFactory points;
points.SetTypeId ("ns3::RandomRectanglePositionAllocator");
points.Set ("X", StringValue
("ns3::UniformRandomVariable[Min=100.0|Max=500.0]"));
points.Set ("Y", StringValue
("ns3::UniformRandomVariable[Min=100.0|Max=1000.0]"));
Ptr<PositionAllocator> waypos = points.Create
()->GetObject<PositionAllocator> ();
waypoint.SetMobilityModel
("ns3::RandomWaypointMobilityModel",
    "Speed", StringValue
("ns3::UniformRandomVariable[Min=0.0|Max=4.0]"),
    "Pause", StringValue
("ns3::UniformRandomVariable[Min=0.0|Max=1.0]"),
    "PositionAllocator", PointerValue (waypos));
waypoint.Install (wifiStaNodes);
```

### RandomWalk2dOutdoorMobilityModel

Before now, we have seen how to simulate user mobility in closed spaces or indoor places, but in reality, users will access wireless or cellular networks outdoors too. In particular, to simulate user mobility outdoors, then we can configure ns-3's RandomWalk2dOutdoorMobilityModel on an ns-3 Wi-Fi or cellular node. For example, to simulate the following scenario, we can use RandomWalk2dOutdoorMobilityModel:

- There are multiple buildings in an area
- Users move outside the buildings only
- Users are restricted from entering the buildings

To use RandomWalk2dOutdoorMobilityModel in simulations, users should remember two important details:

- There should be at least one building in the simulation topology
- Outdoor positions should be defined explicitly based on building boundaries

Now, let's see how to use it in our simulations using the following code snippet:

```
Mobility mobility;
Ptr < Building > building;
building = CreateObject<Building> ();
building->SetBoundaries (Box (0,200,0,100,0,10));
building->SetNRoomsX (1);
building->SetNRoomsY (1);
building->SetNFloors (1);
mobility.
SetMobilityModel("ns3::RandomWalk2dOutdoorMobilityModel",
                           "Bounds", RectangleValue (
                           Rectangle (0,300,0,150)));
  Ptr<OutdoorPositionAllocator> position =
CreateObject<OutdoorPositionAllocator> ();
  Ptr<UniformRandomVariable> xPos =
CreateObject<UniformRandomVariable>();
  xPos->SetAttribute ("Min", DoubleValue (0));
  xPos->SetAttribute ("Max", DoubleValue (300));
  Ptr<UniformRandomVariable> yPos =
CreateObject<UniformRandomVariable>();
  yPos->SetAttribute ("Min", DoubleValue (0));
  yPos->SetAttribute ("Max", DoubleValue (150));
  position->SetAttribute ("X", PointerValue (xPos));
  position->SetAttribute ("Y", PointerValue (yPos));
  mobility.SetPositionAllocator (position);
  mobility.Install (wifiStaNodes);
```

Here, we created a building with boundaries of (0,200,0,100). Then, we set up the outdoor mobility boundaries as (0, 300, 0, 150). Hence, outdoor positions are allocated to users based on these specific boundaries.

That's good. We have explored various ns-3 supporting mobility placement and mobility models for either Wi-Fi or mobile networks. Without delay, we will set up a wireless ad hoc network simulation using ns-3 Wi-Fi PHY and MAC features and placement and mobility models in the next section.

# Setting up wireless ad hoc networks

Setting up wireless ad-hoc networks is highly useful to simulate monitoring an environment using Wi-Fi nodes and offering emergency communication in case of no infrastructure for rescue operations. In ns-3, using various propagation loss, delay, and error models, placement and mobility models, and ad hoc routing algorithms, it is possible to set up realistic ad hoc networks. In this section, we will discuss the following hands-on activity to show how to set up a wireless ad hoc network:

- We assume we want to set up a wireless ad hoc network in a building containing multiple floors and rooms.

- Wi-Fi ad hoc nodes are randomly distributed in rooms of the building.

- All Wi-Fi nodes randomly move inside building rooms and building boundaries only.

- Wi-Fi nodes send data at a constant rate to their gateway node. Two of the Wi-Fi nodes are designated as gateway nodes (or sink nodes).

- Any Wi-Fi ad hoc routing algorithm can be configured.

- At gateway nodes or sink nodes, the receiving bit rate needs to be computed.

- As Wi-Fi nodes are randomly distributed in various places, hidden node issues should be minimized.

- We need to test free space propagation loss and log distance propagation loss model effects in the networks.

- We need to check how changing rate control algorithms will improve the network performance.

## Setting up a wireless or mobile ad hoc network in ns-3

Let's simulate this hands-on in `pkt-adhoc.cc` by taking the following steps using ns-3 code snippet:

1. Start by importing the following necessary ns-3 packages for simulating wireless ad hoc networks:

```
#include <sstream>
#include "ns3/core-module.h"
#include "ns3/network-module.h"
#include "ns3/applications-module.h"
#include "ns3/mobility-module.h"
#include "ns3/buildings-module.h"
#include "ns3/config-store-module.h"
#include "ns3/netanim-module.h"
#include "ns3/internet-module.h"
#include "ns3/yans-wifi-helper.h"
```

```
#include "ns3/flow-monitor-module.h"
using namespace ns3;
Ptr<PacketSink> psink1;
Ptr<PacketSink> psink2;
uint64_t lastTotalRx1 = 0;
uint64_t lastTotalRx2 = 0;
```

Here, we define two global `PacketSink` pointers and variables to compute the throughput at packet sinks in our simulation topology.

2. In `main ()`, begin by creating the necessary number of ns-3 nodes for Wi-Fi ad hoc nodes and sink nodes using the following code snippet. In our simulation, we designate nodes 18 and 19 as sink nodes or gateway nodes:

```
int main (int argc, char *argv[])
{
  uint32_t nSinks = 2;
  double TotalTime = 125.0;
  double dataTime = 120.0;
  double ppers = 1;
  double dataStart = 100.0; // start sending data at 100s
  SeedManager::SetSeed (10);
  SeedManager::SetRun (1);
  NodeContainer adhocNodes1;
  adhocNodes1.Create (18);
  NodeContainer sinkNodes2;
  sinkNodes2.Create (2);
  NodeContainer adhocNodes;
  adhocNodes.Add(adhocNodes1);
  adhocNodes.Add(sinkNodes2);
  NetDeviceContainer allDevices1;
  NetDeviceContainer allDevices2;
```

3. Let's configure the RTS/CTS threshold in our simulation setup to protect transmissions from hidden terminal (or node) transmission issues using the following code. Initially, we comment it during the evaluation:

```
//Config::SetDefault
("ns3::WifiRemoteStationManager::RtsCtsThreshold",
StringValue ("0"))
```

4. Next, deploy two Wi-Fi gateway nodes or sink nodes at locations of (0,0) and (100,100) inside the building boundaries, and configure their mobility as static:

```
MobilityHelper mobility1;
Ptr<ListPositionAllocator> positionAlloc =
CreateObject<ListPositionAllocator> ();
positionAlloc->Add (Vector (0.0, 0.0, 0.0));
positionAlloc->Add (Vector (100.0, 100.0, 0.0));
mobility1.SetPositionAllocator (positionAlloc);
mobility1.SetMobilityModel
("ns3::ConstantPositionMobilityModel");
mobility1.Install(sinkNodes2);
```

5. After deploying the gateway nodes, to deploy Wi-Fi ad hoc nodes, let's create two buildings, each containing two floors, and each floor containing two rooms on the *x*-axis and four rooms on the *y*-axis:

```
Ptr<GridBuildingAllocator>  gridBuildingAllocator;
gridBuildingAllocator =
CreateObject<GridBuildingAllocator> ();
gridBuildingAllocator->SetAttribute ("GridWidth",
UintegerValue (2));
gridBuildingAllocator->SetAttribute ("LengthX",
DoubleValue (50));
gridBuildingAllocator->SetAttribute ("LengthY",
DoubleValue (50));
gridBuildingAllocator->SetAttribute ("DeltaX",
DoubleValue (2));
gridBuildingAllocator->SetAttribute ("DeltaY",
DoubleValue (2));
gridBuildingAllocator->SetAttribute ("Height",
DoubleValue (20));
gridBuildingAllocator->SetBuildingAttribute ("NRoomsX",
UintegerValue (2));
gridBuildingAllocator->SetBuildingAttribute ("NRoomsY",
UintegerValue (4));
gridBuildingAllocator->SetBuildingAttribute ("NFloors",
UintegerValue (2));
gridBuildingAllocator->SetAttribute ("MinX",
DoubleValue (0));
```

```
    gridBuildingAllocator->SetAttribute ("MinY",
DoubleValue (0));
    gridBuildingAllocator->Create (2);
```

6.  Now, randomly deploy all the Wi-Fi nodes inside the building rooms using the following simple code snippet. Restrict users to roam only inside building boundaries:

```
    MobilityHelper mobility;
    Ptr<RandomRoomPositionAllocator> position =
CreateObject<RandomRoomPositionAllocator> ();
    mobility.SetPositionAllocator (position);
    mobility.SetMobilityModel
("ns3::RandomWalk2dMobilityModel",
                              "Mode", StringValue
("Time"),
                              "Time", StringValue ("2s"),
                              "Speed", StringValue
("ns3::ConstantRandomVariable[Constant=1.0]"),
                              "Bounds", RectangleValue
(Rectangle (0.0, 110.0, 0.0, 110.0)));
    mobility.Install (adhocNodes1);
```

By using the following ns-3 code snippets, let's configure all the Wi-Fi ad hoc nodes with the 802.11 a standard, and set up realistic Wi-Fi channel modeling by including ns-3 propagation delay (`ConstantSpeed`), loss (`Friis`), and error models. Then install the MAC and PHY protocols on all ad hoc nodes and collect their network devices for later configurations.

7.  In the following code snippets, note that we allow users to configure suitable ns-3 supporting rate control algorithms through the `argv[1]` command-line argument:

```
    WifiHelper wifi;
    wifi.SetStandard (WIFI_STANDARD_80211a);
    YansWifiPhyHelper wifiPhy;
    YansWifiChannelHelper wifiChannel;
    wifiChannel.SetPropagationDelay
("ns3::ConstantSpeedPropagationDelayModel");
    wifiChannel.AddPropagationLoss
("ns3::LogDistancePropagationLossModel");
    wifiPhy.SetErrorRateModel("ns3::YansErrorRateModel");
    wifiPhy.SetChannel (wifiChannel.Create ());
```

```
WifiMacHelper wifiMac;
wifi.SetRemoteStationManager (argv[1]);
wifiMac.SetType ("ns3::AdhocWifiMac");
allDevices1 = wifi.Install (wifiPhy, wifiMac,
adhocNodes1);
allDevices2 = wifi.Install (wifiPhy, wifiMac,
sinkNodes2);
NetDeviceContainer allDevices
=NetDeviceContainer(allDevices1,allDevices2);
```

8.  Complete our wireless ad hoc network setup by configuring the IP addresses for all nodes using the following code snippet:

```
InternetStackHelper internet;
internet.Install (adhocNodes);
Ipv4AddressHelper address;
address.SetBase ("10.1.1.0", "255.255.255.0");
Ipv4InterfaceContainer allInterfaces;
allInterfaces = address.Assign (allDevices);
```

That's superb. We have completed setting up a wireless ad hoc network suitable for our hands-on activity. Next, we will proceed to test our simulation topology using the following ns-3 traffic applications.

9.  Start by installing packet sink applications on sink nodes 18 and 19 to receive data from various ad hoc nodes using the following lines of code. Here, we can observe that we collect packet sink application pointers in our two global packet sink pointers (psink1 and psink2) to compute receiving throughput at the respective sinks:

```
uint16_t port1 = 10000;
uint16_t port2 = 20000;
double randomStartTime = (1 / ppers) / nSinks;
PacketSinkHelper sink1 ("ns3::UdpSocketFactory",
InetSocketAddress (Ipv4Address::GetAny (), port1));
ApplicationContainer apps_sink1 = sink1.Install
(adhocNodes.Get (18));
psink1 = StaticCast<PacketSink> (apps_sink1.Get (0));
apps_sink1.Start (Seconds (0.0));
apps_sink1.Stop (Seconds (TotalTime));
```

```
    PacketSinkHelper sink2 ("ns3::UdpSocketFactory",
InetSocketAddress (Ipv4Address::GetAny (), port2));
    ApplicationContainer apps_sink2 = sink2.Install
(adhocNodes.Get (19));
    psink2 = StaticCast<PacketSink> (apps_sink2.Get (0));
    apps_sink2.Start (Seconds (0.0));
    apps_sink2.Stop (Seconds (TotalTime));
```

10. Finally, close main () by enabling animation, releasing all simulation resources, and printing the packet sinks' received throughput:

```
    AnimationInterface anim ("adhoc_demo.xml");
    anim.UpdateNodeDescription (18, "Sink-1");
    anim.UpdateNodeDescription (19, "Sink-2");
    anim.EnablePacketMetadata (); // Optional
    anim.EnableIpv4RouteTracking ("adhoc_routes.xml",
Seconds (100), Seconds (125), Seconds (25));
    Simulator::Stop (Seconds (TotalTime));
    Simulator::Run ();
    double averageThroughput1 = ((psink1->GetTotalRx () *
8) / (1e6 * TotalTime));
    double averageThroughput2 = ((psink2->GetTotalRx () *
8) / (1e6 * TotalTime));
    Simulator::Destroy ();
    std::cout << "\nAverage throughput: " <<
averageThroughput1 << " Mbit/s" << std::endl;
    std::cout << "\nAverage throughput: " <<
averageThroughput2 << " Mbit/s" << std::endl;
    }
```

Well done. Our wireless ad hoc network simulation setup is ready for testing.

## Evaluation of the ns-3 wireless ad hoc network

Let's start our hands-on testing activities in the following situations:

- Testing our wireless ad hoc network setup under free space propagation losses
- Testing our wireless ad hoc network setup under realistic propagation losses
- Inspecting a wireless ad hoc network simulation using NetAnim

### Testing our wireless ad-hoc network set up under free space propagation losses

Let's configure the `ConstantSpeedPropagationDelay`, `FriisPropagationLoss`, and `YansError` rate models to simulate free space propagation losses. To test the network performance in terms of the received throughput at the packet sinks, configure the following rate control algorithms: ns-3's `ConstantRate`, `Arf`, and `Minstrel`. Let's execute `pkt-adhoc.cc` from our ns-3 installation directory using the following commands:

```
./ns3 run scratch/pkt-adhoc -- "ns3::ConstantRateWifiManager"
Average throughput at Sink-1: 0.142848 Mbit/s
Average throughput at Sink-2: 0.10304 Mbit/s


/ns3 run scratch/pkt-adhoc -- "ns3::ArfWifiManager"
Average throughput at Sink-1: 0.142848 Mbit/s
Average throughput at Sink-2: 0.089856 Mbit/s


./ns3 run scratch/pkt-adhoc -- "ns3::MinstrelWifiManager"
Average throughput at Sink-1: 0.142848 Mbit/s
Average throughput at Sink-2: 0.132096 Mbit/s
```

From the results, we can observe that change to the `RateControl` algorithm has a significant effect on network performance. For example, by configuring the `MinstrelWifiManager` rate control algorithm, we observed the maximum received throughputs at `Sink-1` and `Sink-2`, and saw that the `ArfWifiManager` configuration led to poor throughput at `Sink-2`.

Let's check whether enabling RTS/CTS improve the network performance with the `ArfWifiManager` rate control algorithm. Uncomment the RTS/CTS enabling lines of code (in `pkt-adhoc.cc`) and run the following command:

```
/ns3 run scratch/pkt-adhoc -- "ns3::ArfWifiManager"
Average throughput at Sink-1: 0.142848 Mbit/s
Average throughput at Sink-2: 0.10304 Mbit/s
```

Oh! That means *hidden terminals* are there in the network and their effect is minimized by channel reservation using an RTS/CTS control message exchange.

That's a good observation, but we ran our simulation in a free space propagation loss model, assuming there is free space between the source and destination Wi-Fi nodes.

### Testing our wireless ad hoc network setup under realistic propagation losses

In order to simulate a real-time wireless environment, let's configure ns-3's `LogDistancePropagation` loss model and observe the network performance results using ns-3's `ConstantRate`, `Arf`, and `Minstrel` rate control algorithms:

```
./ns3 run scratch/pkt-adhoc -- "ns3::ConstantRateWifiManager"
Average throughput at Sink-1: 0.075136 Mbit/s
Average throughput at Sink-2: 0 Mbit/s


/ns3 run scratch/pkt-adhoc -- "ns3::ArfWifiManager"
Average throughput at Sink-1: 0.075136 Mbit/s
Average throughput at Sink-2: 0 Mbit/s


./ns3 run scratch/pkt-adhoc -- "ns3::MinstrelWifiManager"
Average throughput at Sink-1: 0.075136 Mbit/s
Average throughput at Sink-2: 0 Mbit/s
```

When simulating an ad hoc network using the `LogDistancePropagtionLoss` model, changing the rate control algorithm alone is not helping. Let's check whether enabling RTS/CTS helps improve the network performance:

```
./ns3 run scratch/pkt-adhoc -- "ns3::MinstrelWifiManager"
Average throughput at Sink-1: 0.075136 Mbit/s
Average throughput at Sink-2: 0 Mbit/s
```

Oh! There is still no improvement in the throughput at the sinks. Let's inspect what is happening at `Sink-2` using **NetAnim**.

### Inspecting a wireless ad hoc network simulation using NetAnim

Let's check what is happening in our wireless ad hoc network topology (specifically at `sink-2`) by visualizing the simulation execution using NetAnim.

Open the `adhoc-demo.xml` file (generated at the end of the simulation) from **NetAnim** and play the simulation:

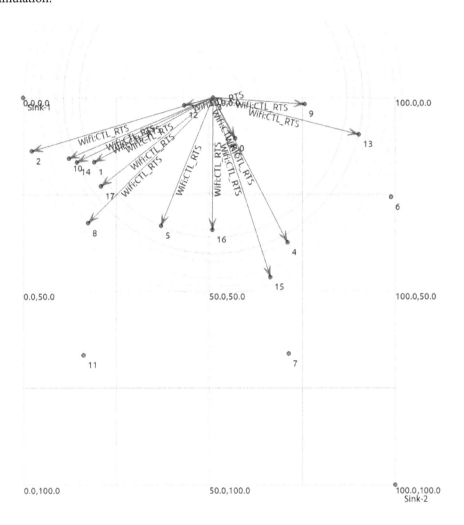

Figure 5.1 – A snapshot of our ad hoc network simulation visualization

When visualizing the simulation (*in Figure 5.1*), we can observe that Wi-Fi nodes are unable to send any packets to `Sink-2`. Hence, we observed **zero throughput** at `Sink-2`. This must be due to a lack of suitable ad hoc routing algorithms.

In this hands-on activity, we did not enable any ad hoc routing algorithms. We evaluated the simulation results using ns-3 default routing: `Ipv4GlobalRouting`. In the next section, we will discuss how to use various ns-3 ad hoc routing algorithms and evaluate simulations.

# Configuring and installing a variety of ad hoc network routing algorithms in ns-3

In the previous section, we discussed how to set up a realistic ad hoc network and test it using the default routing algorithms. In wireless ad hoc networks, the network topology frequently changes. Hence, it is necessary to configure suitable ad hoc routing algorithms for message exchanges between ad hoc Wi-Fi nodes. In the literature, ad hoc routing algorithms are also known as MANET routing algorithms. ns-3 supports the following set of ad hoc routing algorithms to use in ad hoc Wi-Fi network simulations:

- **Proactive routing protocols**: Proactive routing algorithms maintain the routes for every node in the network. Equally, they maintain routing tables with periodic updates. If the network topology changes, broadcast messages are sent to update the routes. Usually, these protocols are suitable for small networks with minimal topology changes. Having routes for every node available at all nodes in the network, messages can be sent between nodes with lower delays. Currently, ns-3 supports two types of proactive routing protocols: the **Destination Sequenced Distance Vector** (**DSDV**) and **Optimized Link State Routing** (**OLSR**) routing protocols.

- Reactive routing protocols: Reactive routing algorithms are also known as on-demand route configuring algorithms. Reactive routing algorithms look for routes between any two nodes only when these nodes need to communicate. Usually, reactive algorithms have route discovery and route maintenance phases. Hence, these routing tables are smaller in size. However, they incur route configuration delays when transmitting any messages. Reactive routing algorithms are suitable for frequent changes in network topologies. Currently, ns-3 supports two types of reactive routing protocols: **Dynamic Source Routing** (**DSR**) and **Ad Hoc On-Demand Distance Vector Routing** (**AODV**).

- ns-3 users can configure these protocols in their simulations with a few lines of code. We will discuss this in an upcoming hands-on activity. For the implementation and working details of these algorithms, we recommend users go through the ns-3 documentation, as well as other mobile ad hoc network routing algorithm resources.

## Configuring and installing various ns-3 ad hoc routing protocols in your simulations

Specifically, we will discuss the following details in a simulation activity:

- We use the `packt-adhoc.cc` source code to configure ns-3's ad hoc routing algorithms (DSDV, OLSR, DSR, and AODV) and evaluate them

- At the end of the simulation, we observe the throughput received at the sink nodes (`Sink-1` and `Sink-2`)

- We see how to use ns-3's energy model to monitor a node's energy consumption details

- We monitor the various routing tables generated by routing algorithms at various nodes

Let's make the following necessary changes to `pkt-adhoc.cc` to configure various ad hoc routing algorithms and monitor the energy consumption details of a sink node. We will save all these changes in a new file called `pkt-adhoc-routing.cc`.

In addition to existing packages in `pkt-adhoc.cc`, include the following necessary ns-3 packages for energy monitoring and the usage of ad hoc routing algorithms:

```
#include "ns3/olsr-module.h"
#include "ns3/dsr-module.h"
#include "ns3/dsdv-module.h"
#include "ns3/aodv-module.h"
#include "ns3/yans-wifi-helper.h"
#include "ns3/flow-monitor-module.h"
#include "ns3/energy-module.h"
#include "ns3/wifi-radio-energy-model-helper.h"
```

After the Wi-Fi PHY and MAC configuration code, include the following lines of code to install the Wi-Fi protocols:

```
allDevices1 = wifi.Install (wifiPhy, wifiMac, adhocNodes1);
allDevices2 = wifi.Install (wifiPhy, wifiMac, sinkNodes2);
NetDeviceContainer allDevices
=NetDeviceContainer(allDevices1,allDevices2);
```

Include the following code snippet for installing ns-3 energy models on Wi-Fi nodes to monitor their energy consumption details:

```
BasicEnergySourceHelper basicSourceHelper;
basicSourceHelper.Set ("BasicEnergySourceInitialEnergyJ",
DoubleValue (1000));
EnergySourceContainer sources = basicSourceHelper.Install
(adhocNodes);
WifiRadioEnergyModelHelper radioEnergyHelper;
radioEnergyHelper.Set ("TxCurrentA", DoubleValue (0.0174));
DeviceEnergyModelContainer deviceModels = radioEnergyHelper.
Install (allDevices, sources);
```

After installing the energy models on the Wi-Fi nodes, install ns-3's various ad hoc routing algorithms, such as DSR, OLSR, DSDV, and AODV, using the following code snippets. Users can select which ad hoc routing to configure on the Wi-Fi nodes by passing the respective algorithm names as the second command-line argument (argv[2]):

```
InternetStackHelper internet;
if (strcmp(argv[2],"dsr")==0)
{
  DsrMainHelper dsrMain;
  DsrHelper dsr;
  internet.Install (adhocNodes);
  dsrMain.Install (dsr, adhocNodes);
  AsciiTraceHelper ascii;
  Ptr<OutputStreamWrapper> stream = ascii.CreateFileStream
("dsrtesting.tr");
  wifiPhy.EnableAsciiAll (stream);


}
else
if (strcmp(argv[2],"dsdv")==0)
{
  DsdvHelper dsdv;
  internet.SetRoutingHelper (dsdv);
  internet.Install (adhocNodes);
  Ptr<OutputStreamWrapper> routingStream =
Create<OutputStreamWrapper> ("dsdv.routes", std::ios::out);
  dsdv.PrintRoutingTableAllAt (Seconds (125), routingStream);
}
else
if (strcmp(argv[2],"aodv")==0)
{
  AodvHelper aodv;
  internet.SetRoutingHelper (aodv);
  internet.Install (adhocNodes);
  Ptr<OutputStreamWrapper> routingStream =
Create<OutputStreamWrapper> ("aodv.routes", std::ios::out);
  aodv.PrintRoutingTableAllAt (Seconds (125), routingStream);
}
```

```
  else
  if (strcmp(argv[2],"olsr")==0)
  {
    OlsrHelper olsr;
    internet.SetRoutingHelper (olsr);
    internet.Install (adhocNodes);
    Ptr<OutputStreamWrapper> routingStream =
Create<OutputStreamWrapper> ("olsr.routes", std::ios::out);
    olsr.PrintRoutingTableAllAt (Seconds (125), routingStream);
  }
```

Then, install the network applications (refer to pkt-adhoc.cc). After installing the network applications, include the following lines of code to monitor Sink-1's (Wi-Fi node 18's) energy consumption traces:

```
Ptr<BasicEnergySource> sinkPtr = DynamicCast<BasicEnergySource>
(sources.Get(18));
  sinkPtr->TraceConnectWithoutContext ("RemainingEnergy",
MakeCallback (&SinkRemainingEnergy));
  Ptr<DeviceEnergyModel> energyPtr =
  sinkPtr->FindDeviceEnergyModels
("ns3::WifiRadioEnergyModel").Get (0);
  sinkPtr->TraceConnectWithoutContext
("TotalEnergyConsumption", MakeCallback (&SinkTotalEnergy));
```

Also, include the following two callback functions for energy monitoring a node (Sink-1) before main() to print the tracing results:

```
void
SinkRemainingEnergy (double oldValue, double remainingEnergy)
{
        std::cout<< Simulator::Now ().GetSeconds ()
                << "s Current remaining energy = " <<
remainingEnergy << "J\n" ;
}
void
SinkTotalEnergy (double oldValue, double totalEnergy)
{
        std::cout<< Simulator::Now ().GetSeconds ()
```

```
                    << "s Total energy consumed by radio = " <<
totalEnergy << "J\n";
}
```

That's all we need to do to evaluate various ns-3-supported routing algorithms and the energy consumption details. So, let's start testing our simulation.

## Evaluating the ns-3 Wi-Fi ad hoc routing algorithms

In the previous hands-on activity, we already confirmed that the LogDistancePropagationLoss model is a more realistic simulation environment, and the Arf rate control algorithm is offering the lowest throughput results. Hence, to check how ns-3's ad hoc routing algorithms will help in these environments, we will start by testing our hands-on activities using the ConstantSpeedPropagationDelay, LogDistancePropagationLoss, and YansError rate models, and the ns-3 Arf rate control algorithm.

### Testing our ad hoc network setup using various ns-3 ad hoc routing algorithms and a basic energy model

We can configure various ns-3 ad hoc routing algorithms and check the throughput at Sink-1 and Sink-2. As we configure monitoring the energy consumption at Sink-1, we can check the remaining energy available at Sink-1 at the end of the simulation.

Let's run the following commands on the ns-3 installation directory. At the end of the simulation, we can observe the routing details of DSDV, AODV, and OLSR in the dsdv.routes, aodv.routes, and olsr.routes files:

```
./ns3 run scratch/pkt-adhoc-routing -- "ns3::ArfWifiManager"
"dsdv"
Average throughput at Sink-1: 0.117632 Mbit/s
Average throughput at Sink-2: 0.085632 Mbit/s
124.997s Current remaining energy = 897.138J


./ns3 run scratch/pkt-adhoc-routing -- "ns3::ArfWifiManager"
"olsr"
Average throughput at Sink-1: 0.119296 Mbit/s
Average throughput at Sink-2: 0.03712 Mbit/s
124.983s Current remaining energy = 897.178J


./ns3 run scratch/pkt-adhoc-routing -- "ns3::ArfWifiManager"
"dsr"
```

```
Average throughput at Sink-1: 0.09664 Mbit/s
Average throughput at Sink-2: 0.06848 Mbit/s
124.981s Current remaining energy = 896.725J


./ns3 run scratch/pkt-adhoc-routing -- "ns3::ArfWifiManager"
"aodv"
Average throughput at Sink-1: 0.08384 Mbit/s
Average throughput at Sink-2: 0.041088 Mbit/s
124.982s Current remaining energy = 897.285J
```

From the results, we can see the throughput at sink nodes 1 and 2, as well as the remaining energy at Sink-1. We recommend that readers modify the energy monitoring code to observe the energy consumption details of various nodes in the network.

In this particular network topology, we can observe that proactive routing algorithms (DSDV and OLSR) perform better compared to reactive algorithms (DSR and AODV). This is due to most of the nodes in our network sending traffic to their sink nodes. With pre-computed routes for sink nodes provided by proactive algorithms, they offer better throughput at sink nodes.

### Inspecting various ad hoc routing algorithm routing tables

First, observe the proactive ad hoc routing algorithm (DSDV and OLSR) routing tables using the following command:

```
cat dsdv.routes
It displays all nodes routing tables. And each node is
maintaining all other nodes routes.
cat olsr.routes
It displays all nodes routing tables. And each node is
maintaining all other nodes routes.
```

As DSDV and OLSR are proactive routing protocols, they maintain the routing details of all nodes.

Similarly, we can check AODV's routing tables using the following command:

```
cat aodv.routes
It displays all nodes routing tables. However, every node
is not maintaining all other nodes routing information.
Especially, we can observe node-6, node-9, node-13, and node-
19. They are not keeping all other nodes routes.
```

As AODV is a reactive routing protocol, it need not maintain all the node routing information at every node.

The current ns-3 implementation does not support capturing DSR routes directly, but we can inspect the DSR routing headers using the `dsrtesting.tr` file, which is generated at the end of the simulation. For example, we can use the following command to view the DSR routing headers:

```
cat dsrtesting.tr |grep 'DsrRoutingHeader'
```

Well done! We have successfully set up realistic wireless ad hoc networks and evaluated the network performance and energy details by configuring ns-3's propagation loss models, error rate models, ad hoc routing algorithms, and energy model.

## Summary

In this chapter, users were introduced to how to set up real-time wireless networks in ns-3 using propagation loss models, error models, and rate control algorithms. We also introduced how to use standard Wi-Fi MAC features for setting up wireless ad hoc networks or LANs, configuring rate control algorithms, handling hidden nodes, and handling large frame transmissions.

In particular, users were introduced to wireless ad hoc network setups through interesting hands-on activities. As part of evaluating ad hoc networks, users practiced how to enable ns-3's ad hoc routing algorithms (DSDV, OLSR, DSR, and AODV), and how to use an ns-3 energy model to estimate the energy consumption of ns-3 Wi-Fi nodes. In the next chapter, we are going to learn about advanced Wi-Fi technology (802.11 n, ac, and ax) feature simulations using interesting ns-3 hands-on activities.

# Researching Advanced Wi-Fi Technologies – 802.11n, ac, and ax in ns-3

In *Chapter 5*, we explored and practiced various ns-3-supported Wi-Fi features for simulating basic wireless networking technologies such as 802.11a, b, and g. In this chapter, we are going to explore and simulate ns-3 supporting advanced wireless networking technologies such as 802.11n, ac, and ax. The ns-3 community is also planning to introduce Wi-Fi 6E in future releases. This chapter mainly discusses how to use the advanced features of 802.11n, ac, and ax in simulations by quickly introducing their high-level details. Hence, we assume you have basic knowledge of the technical details of 802.11n, ac, and ax features before starting this chapter. This chapter starts by introducing various ns-3-supported 802.11n, ac, and ax PHY and MAC features. In the first section of the chapter, we will discuss ns-3 supporting advanced 802.11n, ac, and ax PHY layer features, such as channel bonding, **Orthogonal Frequency-Division Multiplexing (OFDM)**, **Orthogonal Frequency-Division Multiple Access (OFDMA)**, and **Multiple Input Multiple Output (MIMO)**. Then, we will discuss the unique ns-3 MAC features of 802.11n, ac, and ax, such as **Frame Aggregation (FA)**, backward compatibility support, resource scheduling, and interference management.

Later in the second section, you will learn about ns-3 802.11n ac Wi-Fi network simulations. Mainly, we will discuss how to configure and utilize advanced PHY and MAC features, such as channel bonding and FA, to improve channel utilization and offer **Higher Throughput (HT)** for Wi-Fi users. The third section of the chapter discusses, beyond the channel-bonding feature, how 802.11n/ac features such as enabling MIMO and higher **Modulation Coding Schemes (MCSes)** will further improve channel utilization and offer **Very High Throughput (VHT)**.

Finally, in the last two sections, we will introduce the ns-3-supported unique features of 802.11ax, known as Wi-Fi 6 or high-efficiency Wi-Fi networks. Mainly, we will discuss an interesting hands-on activity using the ns-3 Wi-Fi 6 resource scheduling features to support multiple-user parallel transmission and minimize any delay in users' channel access. We will also discuss another interesting hands-on activity related to the **Basic Service Set** (**BSS**) coloring feature of 802.11ax, which maximizes parallel transmissions in dense wireless networks with frequency reuse deployments. Mainly, in this chapter, we will cover the following topics:

- Understanding the 802.11n, ac, and ax Wi-Fi PHY and MAC enhancements supported in ns-3
- Simulating advanced Wi-Fi features – channel bonding and FA
- Simulating advanced Wi-Fi features – MIMO and OFDM
- Exploring spectrum/resource management in Wi-Fi 6 using ns-3 simulations
- Exploring spectrum reuse in Wi-Fi 6 using ns-3 simulations

## Understanding the 802.11n, ac, and ax Wi-Fi PHY and MAC enhancements supported in ns-3

In order to improve wireless spectrum utilization and offer higher bit rates to Wi-Fi users, 802.11n, ac, and ax come with various enhancements at the PHY and MAC layers. In this section, readers will be introduced to ns-3 supporting 802.11n, ac, and ax PHY layer enhancements, and ns-3 MAC layers enhancements related to the 802.11n, ac, and ax standards will also be discussed.

ns-3 802.11n, ac, and ax support the following enhancements:

- Channel bonding
- Only **Single-User MIMO** (**SU-MIMO**)
- **OFDM** and higher MCSes

Besides the aforementioned features, only 802.11ax supports the following:

- OFDMA for resource scheduling
- BSS coloring for dense deployment with frequency reuse 1

Note that ns-3.36 does not support the following enhancements related to the 802.11n, ac, and ax standards:

- 802.11n, ac, and ax beamforming
- 802.11ac and ax **Multi-User MIMO** (**MU-MIMO**)
- Energy consumption modeling for MIMO
- RTS/CTS in the case of channel bonding
- 802.11n **Reduced Inter-Frame Spacing** (**RIFS**)

## ns-3 supporting 802.11n, ac, and ax Wi-Fi PHY layer features

As a first step, the 802.11n and ac standards come with important enhancements such as support for channel bonding, utilizing multiple antennas for transmission or reception, and using higher MCSes at the PHY layer.

### *Channel bonding*

Channel bonding means the ability to combine neighbor channels of the Wi-Fi operating spectrum to enjoy a higher bandwidth. In 802.11n-standard wireless networks without channel bonding, a Wi-Fi node's operating channel bandwidth is limited to 20 MHz. In the case of channel bonding, 802.11n nodes can operate up to 40 MHz bandwidth channels. An ns-3 802.11n implementation supports the following operating channels without channel bonding (i.e., 20 MHz only):

```
36, 40, 44, 48, 52, 56, 60, 64, 100, 104, 108, 112, 116, 120,
124, 128, 132, 136, 140, 144, 149, 153, 161, 165, 169
```

In the case of channel bonding, an ns-3 802.11n simulation can be configured with the following primary channels at a bandwidth of 40 MHz:

```
38, 46, 54, 62, 102, 110, 118, 126, 134, 142, 151, 159
```

In the case of 802.11n, channel bonding is limited to 40 MHz, whereas 802.11ac and ax support channel bonding up to 160 MHz. In ns-3, besides 40 MHz, 802.11ac and ax Wi-Fi simulations can be configured in 80 MHz or 160 MHz with the respective following primary channels:

```
80 MHz: 42, 58, 106, 122, 138, 155
160 MHz: 50, 114
```

Let's see how simple it is to enable channel bonding in ns-3 using the following example code:

```
YansWifiPhyHelper wifiPhy;
wifiPhy.Set ("ChannelSettings", StringValue ("{42, 80,
BAND_5GHZ, 0}"));
```

Here, we configured 80 MHz-b and width channel bonding for either 11ac or ax with a 5 GHz operating frequency. The operating primary channel number is set to 42.

> **Important note**
>
> 802.11n and ac do not support multi-user scheduling on the spectrum, whereas 802.11ax supports OFDMA for multi-user scheduling on the Wi-Fi operating spectrum.

Next, we will discuss another interesting PHY layer enhancement, improving spectrum efficiency and offering an improved throughput to Wi-Fi users using multiple antennas.

## MIMO antennas

In order to support parallel transmissions and receptions at the PHY layer, the 802.11n standard introduced MIMO technology. It enables Wi-Fi STAs and APs to utilize multiple antennas to improve the SNR during parallel transmissions, as well as parallel receptions. Improving the SNR at the sender and receiver enables parallel transmissions and receptions at improved bit rates. However, with the 802.11n standard, parallel transmissions or receptions are limited to **SU-MIMO** only. That means only a pair of Wi-Fi STAs can exchange multiple data streams in parallel, whereas the 802.11ac Wave 2 and 802.11ax standards support **MU-MIMO** in a 5 GHz frequency band. In the case of MU-MIMO, a single Wi-Fi **Station (STA)** can communicate with multiple Wi-Fi STAs in parallel.

Current ns-3 802.11n, ac, and ax implementations support SU-MIMO only. Let's see how to enable the MIMO feature in ns-3 using the following lines of code:

```
YansWifiPhyHelper wifiPhy;
wifiPhy.Set ("Antennas", UintegerValue (nStreams));
wifiPhy.Set ("MaxSupportedTxSpatialStreams", UintegerValue
(nStreams));
wifiPhy.Set ("MaxSupportedRxSpatialStreams", UintegerValue
(nStreams));
```

In the current ns-3 implementation, we can configure a maximum of four antennas with four Tx spatial streams and four Rx spatial streams (4x4 SU-MIMO).

> **Important note**
> ns-3.36 does not support MU-MIMO.

## OFDM and higher MCSes

By adopting OFDM modulation techniques at the PHY layer, the 802.11n, ac, and ax standards offer resilient PHY layer transmissions, and promise HT and VHT transmissions respectively using the following MCSes for transmissions:

- 802.11n supports HT using the following MCSes: MCS-0 to MCS-7
- 802.11ac supports VHT using the following MCSes: MCS-0 to MCS-8
- 802.11ax supports the following MCSes: MCS-0 to MCS-11

Configuring higher MCSes allows senders to transmit a higher number of bits per symbol in a signal. However, successful reception at the receiver depends on the receiver's SNR or SINR. Therefore, only data signals should be transmitted at higher MCSes, and control signals should be transmitted at lower MCS values. Hence, enabling MIMO along with higher MCSes may help successful receptions at receivers. Let's see how to configure specific MCSes without any rate control algorithms in ns-3 Wi-Fi simulations using the following lines of code:

```
wifi.SetRemoteStationManager ("ns3::ConstantRateWifiManager",
"DataMode",StringValue("HtMcs6"),"ControlMode",
StringValue("HtMcs0"));
```

Here, we configured `ConstantRateWifiManger` to transmit data signals at a higher MCS (`HtMcs6`) and the control signal at a lower MCS (`HtMcs0`).

> **Important note**
>
> Wi-Fi rate control algorithms running at APs or STAs configure their transmissions' MCS values based on channel conditions. Hence, to configure constant MCS values for transmissions, we need to configure `ConstantRateWifiManager` for STAs or APs.

Now, we have discussed ns-3 supporting PHY enhancements useful for improving channel utilization and STA throughput. In the next section, we will discuss ns-3 supporting MAC enhancements related to the 802.11n, ac, and ax standards.

## ns-3 supporting 802.11n, ac, and ax Wi-Fi MAC layer features

In ns-3 802.11n, ac, and ax implementations, the following major MAC enhancements are supported:

- FA in 802.11n, ac, and ax
- Backward compatibility in 802.11n, ac, and ax
- Multi-user scheduling in 802.11ax
- BSS coloring in 802.11ax

### FA in 802.11n, ac, and ax

Besides the PHY layer enhancements, 802.11n also introduced significant MAC enhancements to improve channel utilization by reducing the wait time for channel access, as well as avoiding unnecessary accesses and transmissions. Mainly, 802.11n introduced **FA schemes** to improve channel utilization. Wi-Fi STA channel access involves overhead such as inter-frame spacing, random backoff based on the **Contention Window** (**CW**), and **Acknowledgment** (**ACK**) wait durations for any single frame transmission. In the case of multiple frame transmissions in a flow, the overhead can be minimized to improve channel utilization and throughput for wireless STAs. The 802.11n, ac, and ax standards proposed FA schemes as part of their MAC layer enhancements. By enabling FA schemes, STAs can send multiple frames belonging to a single access category and destined for a single station in a single channel access with the aggregation of multiple frames. Hence, this eliminates overhead, such as the contention overhead, channel access wait times, transmission of redundant radio preamble, and **Media Access Control (MAC)** headers.

802.11n, ac, and ax support the following FA techniques:

- **Aggregate MAC Service Data Unit (A-MSDU)**: With A-MSDU FA, based on the **Aggregation Size (AGG_SIZE)** configured by the MAC, multiple MSDUs are aggregated into a single A-MSDU, as shown in the following figure:

Figure 6.1 – An A-MSDU

- **Aggregate MAC Protocol Data Unit (A-MPDU)**: With A-MPDU FA, based on the AGG_SIZE configured at MAC, multiple 802.11 frames belonging to the same access category and for the same destination are aggregated into a single A-MPDU, as shown in the following figure:

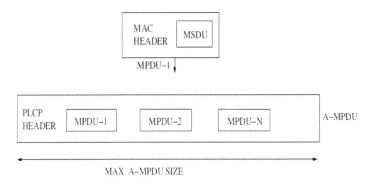

Figure 6.2 – An A-MPDU

Two-level FA

It is also possible to enable two-level aggregation schemes, where multiple A-MSDUs can be aggregated into a single A-MPDU. ns-3 also supports two-level FA.

Let's see how to enable an A-MSDU at an AP in ns-3 simulations using the following lines of code:

```
Ptr<NetDevice> apdev = APs.Get (0)->GetDevice (0);
Ptr<WifiNetDevice> ap_wifi = DynamicCast<WifiNetDevice>
(apdev);
ap_wifi->GetMac ()->SetAttribute ("BE_MaxAmsduSize",
UintegerValue (7935));
```

Let's see how to enable an A-MPDU at an AP in ns-3 simulations using the following lines of code:

```
ap_wifi->GetMac ()->SetAttribute ("BE_MaxAmpduSize",
UintegerValue (32768));
```

Let's see how to enable two-level FA at an AP in ns-3 simulations using the following lines of code:

```
ap_wifi->GetMac ()->SetAttribute ("BE_MaxAmsduSize",
UintegerValue (7935));
ap_wifi->GetMac ()->SetAttribute ("BE_MaxAmpduSize",
UintegerValue (32768));
```

Here, with all FA schemes, we configure the AP with the maximum possible A-MSDU and A-MPDU sizes. We can also configure aggregation sizes lesser than these values. Similarly, we need to enable the respective FA at the STAs.

## Backward compatibility in 802.11n, ac, and ax

In ns-3 Wi-Fi simulations, by default, backward compatibility is supported for the 802.11n, ac, and ax standards. For example, users can simulate mixed wireless STA network scenarios, such as an 802.11ac AP communicating with 802.11n, a, and b STAs. Users need not configure anything specifically to enable backward compatibility. However, Wi-Fi APs or STAs should not be configured with PHY or MAC enhancement features such as channel bonding, MIMO, FA, higher HT/VHT/HE MCS values, or OFDMA.

## Multi-user scheduling in 802.11ax

802.11ax is the first standard to support multi-user scheduling on the Wi-Fi spectrum. That means that in 802.11a, b, g, n, and ac, the entire Wi-Fi operating channel bandwidth is assigned to a single user only after following the channel access rules. The evolution of 802.11ax brings resource scheduling to enable multi-user parallel transmissions on the Wi-Fi spectrum. With the 802.11ax standard, by adopting the OFDMA technology at the PHY layer spectrum, the bandwidth is organized into multiple **Resource Units** (**RUs**). Hence, in a given unit of time, multiple users can be assigned various RUs based on the 802.11ax standard sub-channel assignment rules.

OFDMA offers efficient use of the spectrum by allowing simultaneous transmissions. It minimizes the contention overhead for MAC and channel access delays. Hence, Wi-Fi APs can allocate the spectrum to a single user or multiple users simultaneously based on its STAs' traffic requirements. However, **Carrier-Sense Multiple Access with Collision Avoidance** (**CSMA/CA**) rules are still applicable to the AP. Once the AP has a **transmission opportunity** (**TXOP**) then, only the AP can schedule and control multiple 802.11ax STAs for downlink or uplink transmissions. For example, using OFDMA, the 20 MHz operating spectrum can be divided into 9 smaller subchannels, and then 802.11ax AP can allocate 9 RUs to 9 802.11 11ax STAs simultaneously for small frame transmissions.

Unsurprisingly, our ns-3 802.11ax implementation also supports multi-user scheduling in 802.11ax network simulations. In order to enable resource scheduling in an ns-3 simulation, besides configuring the 802.11ax PHY/MAC with Wi-Fi nodes, we need to use the following ns-3 lines of code to enable resource scheduling using a Round Robin scheduler:

```
WifiMacHelper wifiMac;
wifiMac.SetMultiUserScheduler ("ns3::RrMultiUserScheduler",
                    "EnableUlOfdma", BooleanValue
(true),
                    "EnableBsrp", BooleanValue (true));
```

### BSS coloring in 802.11ax

In order to increase spectrum utilization capacity in a dense deployment of Wi-Fi networks, frequency reuse must be adopted between the BSS. 802.11ax/Wi-Fi 6 comes with a BSS coloring feature, which helps to increase channel reuse. Wi-Fi nodes must follow the channel access rules defined by the **CSMA/CA** protocol to minimize collisions during transmissions. In general, by following the CSMA/CA rules, 802.11 Wi-Fi nodes defer their transmissions based on a **Signal Detect** (**SD**) threshold, for example, a SD threshold greater than or equal to 4 dB or higher than the noise floor means there may be an ongoing transmission. Due to this rule, often, neighboring BSS STAs or APs may defer their transmissions unnecessarily.

For example, BSS-1 and BSS-2 are operating on the same channel-1, if BSS-1's AP or STAs listen to the ongoing transmission from a neighbor AP or STAs belonging to BSS-2, then the BSS-1 STAs or AP defer their transmissions unnecessarily. This problem is due to the overlapping of BSSes, with two BSSes on the same channel that can listen to each other.

With the Wi-Fi 6 standard, this problem is addressed by BSS coloring. The BSS color is an identifier for the BSS. Now, Wi-Fi 6 networks can define the following rules to improve frequency reuse and increase parallel transmissions based on BSS identities (BSS colors):

- Intra-BSS: If the BSS color is the same for the transmitter and listener, then it is considered an intra-BSS transmission, and the listening Wi-Fi nodes will defer its transmission

- Inter-BSS: If the BSS color is different between the transmitter and listener, as the transmitting Wi-Fi node belongs to a different BSS, a listening Wi-Fi node may not defer its transmission

In general, 802.11 Wi-Fi nodes use two separate **Clear Channel Assessment (CCA)** thresholds when listening to the wireless medium. The SD threshold is used to identify the incoming 802.11 preamble transmission from another transmitting 802.11 Wi-Fi node. The **Energy Detect (ED)** threshold is used to detect non-802.11 wireless node transmissions during the CCA. Based on the BSS color detection and using the SD and ED thresholds, Wi-Fi 6 nodes can use **OBSS Power Detection (OBSSPD)** algorithms to dynamically decide and configure the suitable CCA SD thresholds for inter-BSS and intra-BSS transmission rules.

An ns-3 802.11ax implementation supports the following OBSSPD algorithm for simulations. We can configure it using the following lines of code:

```
wifi.SetObssPdAlgorithm ("ns3::ConstantObssPdAlgorithm",
                         "ObssPdLevel", DoubleValue
(obssPdThreshold));
```

Here, `obssPdThreshold` is nothing but the SD threshold.

Now, we have discussed various ns-3 supporting PHY and MAC enhancement features for the 802.11n, ac, and ax standards. In the upcoming sections, we will discuss interesting hands-on activities to simulate ns-3 supporting 802.11n, ac, and ax features.

## Simulating advanced Wi-Fi features – channel bonding and FA

In this section, we discuss how to use and configure all the supported features at the PHY and MAC layers for the advanced Wi-Fi 802.11n and ac standards, such as channel bonding, FA, SU-MIMO, and OFDM, supporting higher MCS values. To study these features systematically in ns-3 simulations, we will discuss a hands-on activity in a step-by-step procedure to configure an 802.11n/ac-supporting Wi-Fi LAN. However, we will first evaluate the 802.11n network only. Specially, we will discuss how the channel-bonding and FA features of the 802.11n standard are helpful to improve the throughput for STAs. In this hands-on activity, we discuss the following important activities:

- Setting up an 802.11n/ac-standard-configurable Wi-Fi LAN with 1 AP and 10 STAs.
- Deploying the AP at location `100,100`.
- Randomly deploying Wi-Fi STAs around the AP within a radius of 30 meters.
- Configuring the `RandomWalk` mobility model for all STAs.
- Configuring the 802.11n/ac PHY level with a 5 GHz operating frequency band and suitable channels to enable channel bonding.

- Configuring the PHY layer for SU-MIMO parallel transmission streams, as well as reception streams. 802.11n and ac both support SU-MIMO, but we will test the SU-MIMO feature in the next hands-on activity in relation to the 802.11ac standard.

- In order to use OFDM and support higher MCS values in our simulations in the next hands-on activity, configuring Wi-Fi MAC with `ConstantRateWifiManger` to test various 802.11ac-supporting MCS values. In this hands-on activity, we do not test the impact of the OFDM MCS values on the throughput for the STAs.

- Configuring Wi-Fi STAs and APs with two-level FA schemes with the maximum possible aggregation sizes.

- Testing a 802.11n-supporting channel-bonding configuration (40 MHz).

- Testing how FA alone improves the throughput for STAs in an 802.11n network.

- Testing how channel bonding and FA improve the throughput for STAs in the 802.11n network together.

## Setting up an 802.11n Wi-Fi LAN in ns-3 for simulating channel bonding and FA

Let's look at the steps to do this:

1. Prepare `pkt-11n-11ac-test.cc` for simulating the hands-on activity by importing all the following necessary ns-3 packages:

```
#include "ns3/core-module.h"
#include "ns3/yans-wifi-helper.h"
#include "ns3/ssid.h"
#include "ns3/mobility-helper.h"
#include "ns3/internet-module.h"
#include "ns3/applications-module.h"
#include "ns3/internet-apps-module.h"
#include "ns3/ipv4-address-helper.h"
#include "ns3/yans-wifi-channel.h"
#include "ns3/wifi-net-device.h"
#include "ns3/wifi-mac.h"
#include "ns3/flow-monitor-module.h"
#include "stdlib.h"
using namespace ns3;
```

2.  Start `main()` by creating the necessary number of ns-3 nodes for setting up an 802.11n/ac Wi-Fi AP and STAs:

```
int main (int argc, char *argv[])
{
  uint32_t nAPs = 1;
  uint32_t nStas = 10;
  NodeContainer APs;
  APs.Create (nAPs);
  NodeContainer stas;
  stas.Create (nStas);
  NetDeviceContainer staDev;
  NetDeviceContainer apDev;
```

3.  Deploy the AP and STAs as per the hands-on activity requirements using the following ns-3 placement and mobility models:

```
MobilityHelper mobility;
Ptr<ListPositionAllocator> positionAlloc =
CreateObject<ListPositionAllocator> ();
  positionAlloc->Add (Vector (100, 100, 0.0));
  mobility.SetPositionAllocator (positionAlloc);
  mobility.SetMobilityModel
("ns3::ConstantPositionMobilityModel");
  mobility.Install (APs.Get (0));
  mobility.SetPositionAllocator
("ns3::RandomDiscPositionAllocator",
                          "X", StringValue ("100.0"),
                          "Y", StringValue ("100.0"),
                          "Rho", StringValue
("ns3::UniformRandomVariable[Min=0|Max=30]"));
  mobility.SetMobilityModel
("ns3::RandomWalk2dMobilityModel",
                          "Mode", StringValue
("Time"),
                          "Time", StringValue
("100s"),
                          "Speed", StringValue
("ns3::ConstantRandomVariable[Constant=1.0]"),
```

```
                                    "Bounds", StringValue
("0|200|0|200"));
    mobility.Install (stas);
```

4.  Next, configure the 802.11n Wi-Fi standard, and create an ns-3 default Wi-Fi channel using the following ns-3 lines of code:

```
WifiHelper wifi;
wifi.SetStandard (WIFI_STANDARD_80211n);
YansWifiPhyHelper wifiPhy;
YansWifiChannelHelper wifiChannel =
YansWifiChannelHelper::Default ();
wifiPhy.SetChannel (wifiChannel.Create ());
```

5.  Next, configure possible 20 MHz channels and the 802.11n/ac-supporting channel-bonding features. However, it is an 802.11n simulation, so, during testing, users should pass either 20 or 40 only as command-line argument 1. Otherwise, it will give errors:

```
int ch_width = atoi(argv[1]);
switch(ch_width)
{
  case 20:
    wifiPhy.Set ("ChannelSettings", StringValue ("{36,
20, BAND_5GHZ, 0}"));
    break;
  case 40:
    wifiPhy.Set ("ChannelSettings", StringValue ("{38,
40, BAND_5GHZ, 0}"));
    break;
  case 80:
    wifiPhy.Set ("ChannelSettings", StringValue ("{42,
80, BAND_5GHZ, 0}"));
    break;
  case 160:
    wifiPhy.Set ("ChannelSettings", StringValue ("{50,
160, BAND_5GHZ, 0}"));
    break;
}
```

6.  After the channel-bonding configuration, include the 802.11n/ac-supporting SU-MIMO configuration in the simulation setup. Allow users to configure the maximum number of antennas and Tx and Rx streams through command-line argument 2. However, we will postpone discussing how to test the MIMO features until the next hands-on activity in regard to 802.11ac. Hence, comment the following lines of code:

```
/*uint32_t nStreams = atoi(argv[2]);
wifiPhy.Set ("Antennas", UintegerValue (nStreams));
  wifiPhy.Set ("MaxSupportedTxSpatialStreams",
UintegerValue (nStreams));
  wifiPhy.Set ("MaxSupportedRxSpatialStreams",
UintegerValue (nStreams));*/
```

Now, we have completed our simulation setup with the advanced features of an 802.11n/ac PHY configuration, such as channel bonding and MIMO.

7.  Next, we will configure the 802.11n MAC and the advanced feature of two-level FA with the maximum possible aggregation size at the AP. During testing, users can either enable or disable FA using command-line argument 2. By passing 0 as an argument, FA is disabled, and passing 1 as an argument enables FA:

```
WifiMacHelper wifiMac;
Ssid ssid = Ssid ("Packt");
wifiMac.SetType ("ns3::ApWifiMac",
                  "Ssid", SsidValue (ssid));
apDev = wifi.Install (wifiPhy, wifiMac, APs.Get (0));
int aggr=atoi(argv[2]);
if (aggr)
{
   Ptr<NetDevice> apdev = APs.Get (0)->GetDevice (0);
   Ptr<WifiNetDevice> ap_wifi =
DynamicCast<WifiNetDevice> (apdev);
   ap_wifi->GetMac ()->SetAttribute ("BE_MaxAmsduSize",
UintegerValue (7935));
   ap_wifi->GetMac ()->SetAttribute ("BE_MaxAmpduSize",
UintegerValue (32768));
 }
```

8. In order to enable FA for a Wi-Fi LAN, it is necessary to enable the same FA scheme at both the AP and STAs. Hence, we enable two-level FA at the STAs using the following code snippet:

```
wifiMac.SetType ("ns3::StaWifiMac",
                "Ssid", SsidValue (ssid));
staDev = wifi.Install (wifiPhy, wifiMac, stas);
if (aggr)
{
   Ptr<NetDevice> stadev = stas.Get (9)->GetDevice (0);
   Ptr<WifiNetDevice> sta_wifi =
DynamicCast<WifiNetDevice> (stadev);
   sta_wifi->GetMac ()->SetAttribute ("BE_MaxAmsduSize",
UintegerValue (7935));
   sta_wifi->GetMac ()->SetAttribute ("BE_MaxAmpduSize",
UintegerValue (32768));
}
```

9. Next, let's see how to use and configure OFDM to support higher MCS values in our simulation setup using the following code. Note that usually, MCS values are decided by rate control algorithms at the Wi-Fi STAs. Hence, to explicitly configure specific MCS values, we need to use ConstantRateWifiManger and configure the suitable MCS values for data signals and control signals separately using the following lines of code. During testing, users can configure the suitable MCS values using command-line argument 3. We also comment the following lines because we will test VHT MCSes in the next hands-on activity alone:

```
/*wifi.SetRemoteStationManager ("ns3::ConstantRateWifi
Manager","DataMode",StringValue(argv[3]),
                                        "ControlMode",
StringValue("VhtMcs0"));*/
```

Well done. We have configured all the advanced 802.11n/ac features in our simulation.

10. Let's install InternetStackHelper and configure the IP addresses for the Wi-Fi AP and its STA using the following code snippet:

```
Ipv4InterfaceContainer staInterface;
Ipv4InterfaceContainer apInterface;
InternetStackHelper stack;
stack.Install(APs);
stack.Install (stas);
Ipv4AddressHelper ip;
ip.SetBase ("192.168.1.0", "255.255.255.0");
```

```
apInterface = ip.Assign (apDev);
staInterface = ip.Assign (staDev);
```

11. Finally, to test the impact of the channel-bonding and FA features on the throughput for the STAs in an 802.11n network, we generate TCP-based constant data rate traffic between a Wi-Fi STA-9 and AP. Here, the AP is sending TCP data at a rate of 200 Mbps using ns-3's OnOffApplication and the ns-3 PacketSink application installed at STA-9:

```
uint64_t port = 9;
Address localAddress (InetSocketAddress
(Ipv4Address::GetAny (), port));
PacketSinkHelper sink ("ns3::TcpSocketFactory",
localAddress);
ApplicationContainer apps_sink = sink.Install (stas.Get
(9));
OnOffHelper onoff ("ns3::TcpSocketFactory",
InetSocketAddress (staInterface.GetAddress(9), port));
onoff.SetConstantRate (DataRate ("200Mbps"), 1000);
onoff.SetAttribute ("StartTime", TimeValue (Seconds
(6)));
onoff.SetAttribute ("OnTime",  StringValue
("ns3::ConstantRandomVariable[Constant=1]"));
onoff.SetAttribute ("OffTime", StringValue
("ns3::ConstantRandomVariable[Constant=0]"));
onoff.SetAttribute ("StopTime", TimeValue (Seconds
(8)));
ApplicationContainer apps_source = onoff.Install (APs.
Get (0));
apps_sink.Start (Seconds (0.5));
apps_sink.Stop (Seconds (10));
```

12. After writing the testing code, close main () by freeing all simulation resources and configure a flow monitor to analyze the simulation results using the following code snippet:

```
Ptr<FlowMonitor> flowmon;
FlowMonitorHelper flowmonHelper;
flowmon = flowmonHelper.InstallAll ();
Simulator::Stop (Seconds(11.0));
Simulator::Run ();
flowmon->SerializeToXmlFile ("80211nspeed.xml", true,
true);
```

```
        Simulator::Destroy ();
    }
```

That's superb. Now, we are ready to test our 802.11n setup and evaluate the channel-bonding and FA features.

> **Important note**
>
> Please note that we commented the SU-MIMO and VHT MCS configuration-related lines of code to test these features in the next hands-on activity in regard to 802.11ac networks.

Let's start testing ns-3 supporting 802.11n features using our simulation setup thoroughly in the following topic.

## Testing the channel-bonding and FA features of 802.11n in an ns-3 simulation

We will first evaluate the channel bonding feature of the 802.11n standard using our simulation setup.

### Testing the channel bonding feature

Let's start testing our simulation setup without channel bonding and FA using the following commands in the ns-3 installation folder. We pass 20 and 0 as command-line arguments 1 and 2 respectively to test 20 MHz and disable FA:

```
./ns3 run scratch/pkt-11n-11ac-test -- 20 0
python3 scratch/flowmon-parse-results.py 80211nspeed.xml
FlowID: 1 (TCP 192.168.1.1/49153 --> 192.168.1.11/9)
TX bitrate: 58061.92 kbit/s
RX bitrate: 58151.36 kbit/s
Mean Delay: 13.54 ms
Packet Loss Ratio: 0.00 %
```

From the results, we can observe that although the AP sends application traffic at 200 Mbps, and STA-9 experiences a throughput of only 59 Mbs. Can we improve STA-9 throughput by enabling channel bonding alone? Let's execute the following commands and check the results:

```
./ns3 run scratch/pkt-11n-11ac-test -- 40 0
python3 scratch/flowmon-parse-results.py 80211nspeed.xml
FlowID: 1 (TCP 192.168.1.1/49153 --> 192.168.1.11/9)
TX bitrate: 122163.56 kbit/s
```

```
RX bitrate: 122351.62 kbit/s
Mean Delay: 5.73 ms
Packet Loss Ratio: 0.00 %
```

Oh, that's great. We can observe STA-9 now experiencing almost double the throughput, 122 Mbps with 40 MHz, of no channel bonding (20 MHz). Next, we will test the 802.11n FA feature.

### Testing the FA feature without channel bonding in our simulation setup

Let's test our simulation setup with a 20 MHz bandwidth and FA features using the following command:

```
./ns3 run scratch/pkt-11n-11ac-test -- 20 1
python3 scratch/flowmon-parse-results.py 80211nspeed.xml
FlowID: 1 (TCP 192.168.1.1/49153 --> 192.168.1.11/9)
TX bitrate: 60733.13 kbit/s
RX bitrate: 60827.10 kbit/s
Mean Delay: 13.74 ms
Packet Loss Ratio: 0.00 %
```

That's good. There is a significant improvement in the throughput for STA-9 (60Mbps). FA helps STA-9 to get 2 Mbps HT compared to no FA.

### Testing channel bonding and FA together in our simulation setup

Finally, let's test whether using channel bonding (40 MHz) and FA together can improve the STA's throughput using the following commands:

```
./ns3 run scratch/pkt-11n-11ac-test -- 40 1
python3 scratch/flowmon-parse-results.py 80211nspeed.xml
FlowID: 1 (TCP 192.168.1.1/49153 --> 192.168.1.11/9)
TX bitrate: 127243.50 kbit/s
RX bitrate: 127436.63 kbit/s
Mean Delay: 5.91 ms
Packet Loss Ratio: 0.00 %
```

From the results, it is clear that enabling channel bonding (40 MHz) and FA on 802.11n Wi-Fi networks improves the throughput for STAs. That's good work. Let's start exploring the ns-3-supported 802.11ac features in the next section.

> **802.11n – MIMO and OFDM support**
>
> Configuring MIMO and HT MCSes is supported in ns-3 802.11n too. In this hands-on activity, we only evaluated the channel bonding and FA features for the 802.11n standard. We recommend you uncomment the SU-MIMO code and the MCS configuration-related code to test this hands-on activity as an exercise. During the evaluation, you will observe that these features further improve 802.11n users' throughput.

## Simulating advanced Wi-Fi features – MIMO and OFDM

In this section, we will focus on an 802.11ac simulation setup and evaluation. As part of 802.11ac, we will configure the advanced Wi-Fi features of MIMO and OFDM supporting VHT MCS values in ns-3 simulations to improve STA throughput. In particular, we will discuss how enabling MIMO helps to improve throughput by using multiple antennas to handle parallel transmissions or reception. Besides transmitting signals at VHT MCSes, the receivers should also receive at good SNR. In these cases, we discuss how enabling MIMO helps transmitters to use HT MCS values. Specifically, we will do the following activities to test the MIMO and OFDM features for an 802.11ac Wi-Fi LAN:

- Using the `pkt-11n-ac-test.cc` source code and making the following necessary changes to simulate 802.11ac networks

- Configuring 802.11ac as the Wi-Fi standard

- Uncommenting the lines of code related to SU-MIMO configuration in `pkt-11n-11ac-test.cc`

- Uncommenting the lines of code related to VHT MCS configuration in `pkt-11n-11ac-test.cc`

- Testing how SU-MIMO parallel Tx/Rx streams improve the throughput between an AP and STAs in our test setup

- Testing how using higher MCSes of 802.11ac improves the throughput between an AP and STAs in our test setup

- Testing how MIMO helps to use higher MCSes of 802.11ac and improve the throughput between an AP and STAs in our test setup

### Setting up and evaluating an 802.11ac Wi-Fi LAN in ns-3 for testing the MIMO and OFDM features

Let's open `pkt-11n-11ac-test.cc` and make the following changes for implementing the 802.11ac-related hands-on activities.

Configure the 802.11ac standard for building VHT Wi-Fi LANs by changing the following line of code in `pkt-11n-11ac-test.cc`:

```
WifiHelper wifi;
wifi.SetStandard (WIFI_STANDARD_80211ac);
```

Next, uncomment the SU-MIMO configuration-related code in our simulation setup. Allow users to configure the maximum number of antennas and Tx and Rx streams through command-line argument 2:

```
uint32_t nStreams = atoi(argv[2]);
wifiPhy.Set ("Antennas", UintegerValue (nStreams));
wifiPhy.Set ("MaxSupportedTxSpatialStreams", UintegerValue
(nStreams));
wifiPhy.Set ("MaxSupportedRxSpatialStreams", UintegerValue
(nStreams));
```

Then, uncomment the VHT MCS configuration-related code in `pkt-11n-11ac-test.cc`. During testing users, can configure the suitable MCS values using command-line argument 3:

```
wifi.SetRemoteStationManager ("ns3::ConstantRateWifiManager",
"DataMode",StringValue(argv[3]),
                                        "ControlMode",
StringValue("VhtMcs0"));
```

The final change is changing the `OnOffApplication` constant data rate to 800 Mbps to test 802.11ac's data rates:

```
onoff.SetConstantRate (DataRate ("800Mbps"), 1000);
```

After making all these changes in `pkt-11ac-test.cc`, let's see how to evaluate it thoroughly in the next topic.

## Testing the MIMO and OFDM features of 802.11ac in an ns-3 simulation

Let's start testing our simulation setup with the maximum possible channel width, which is 160 MHz in the 802.11ac standard (comment the MCS configuration-related lines of code and run the following commands).

### Testing the MIMO features

Start without MIMO and test our simulation setup using the following command:

```
./ns3 run scratch/pkt-11n-11ac-test -- 160 1
python3 scratch/flowmon-parse-results.py 80211acspeed.xml
```

```
FlowID: 1 (TCP 192.168.1.1/49153 --> 192.168.1.11/9)
TX bitrate: 356885.32 kbit/s
RX bitrate: 357425.88 kbit/s
Mean Delay: 1.57 ms
Packet Loss Ratio: 0.00 %
```

We can observe that the AP Tx throughput is only 350 Mbps and the STA-9 RX throughput is only 350 Mbps. Still, the STA's throughput is not reaching to maximum possible throughput. Let's check our simulation results by enabling MIMO with two antennas and two parallel streams:

```
./ns3 run scratch/pkt-11n-11ac-test -- 160 2
python3 scratch/flowmon-parse-results.py 80211acspeed.xml
FlowID: 1 (TCP 192.168.1.1/49153 --> 192.168.1.11/9)
TX bitrate: 432822.76 kbit/s
RX bitrate: 433479.12 kbit/s
Mean Delay: 0.54 ms
```

Oh, that's good. STA-9 is able to receive at a speed of 433 Mbps using 2x2 SU-MIMO.

Let's configure the maximum possible number of antennas and streams to 4 in our simulation setup and test it with the following commands:

```
./ns3 run scratch/pkt-11n-11ac-test -- 160 4
python3 scratch/flowmon-parse-results.py 80211acspeed.xml
FlowID: 1 (TCP 192.168.1.1/49153 --> 192.168.1.11/9)
TX bitrate: 609000.27 kbit/s
RX bitrate: 609938.85 kbit/s
Mean Delay: 1.05 ms
Packet Loss Ratio: 0.00 %
```

That's excellent. STA-9's throughput reaches 600 Mbps using 4x4 SU-MIMO.

### Testing the VHT MCS features

Assuming we don't have MIMO features in our simulation topology, can Wi-Fi STAs get better throughput by using VHT MCS values? Let's test that using the following commands. Start with transmitting data at VhtMcs4:

```
./ns3 run scratch/pkt-11n-11ac-test -- 160 1 VhtMcs4
python3 scratch/flowmon-parse-results.py 80211acspeed.xml
```

```
FlowID: 1 (TCP 192.168.1.1/49153 --> 192.168.1.11/9)
TX bitrate: 210698.97 kbit/s
RX bitrate: 211024.51 kbit/s
Mean Delay: 3.67 ms
Packet Loss Ratio: 0.00 %
```

Oh, that's good – using VhtMcs4 provides better throughput.

Try to test our simulation topology with other supporting VHT MCS values, such as VhtMcs5 to VhtMcs9, by using the following command:

```
./ns3 run scratch/pkt-11n-11ac-test -- 160 1 VhtMcs5
python3 scratch/flowmon-parse-results.py 80211acspeed.xml
FlowID: 1 (TCP 192.168.1.1/49153 --> 192.168.1.11/9)
TX bitrate: 260157.49 kbit/s
RX bitrate: 260559.40 kbit/s
Mean Delay: 2.95 ms
Packet Loss Ratio: 0.00


./ns3 run scratch/pkt-11n-11ac-test -- 160 1 VhtMcs6
python3 scratch/flowmon-parse-results.py 80211acspeed.xml
FlowID: 1 (TCP 192.168.1.1/49153 --> 192.168.1.11/9)
TX bitrate: 283027.89 kbit/s
RX bitrate: 283454.89 kbit/s
Mean Delay: 2.68 ms
Packet Loss Ratio: 0.00 %


./ns3 run scratch/pkt-11n-11ac-test -- 160 1 VhtMcs7
python3 scratch/flowmon-parse-results.py 80211acspeed.xml


FlowID: 1 (TCP 192.168.1.1/49153 --> 192.168.1.11/9)
TX bitrate: 302726.03 kbit/s
RX bitrate: 303186.50 kbit/s
Mean Delay: 2.49 ms
Packet Loss Ratio: 0.00 %


./ns3 run scratch/pkt-11n-11ac-test -- 160 1 VhtMcs8
python3 scratch/flowmon-parse-results.py 80211acspeed.xml
```

```
FlowID: 1 (TCP 192.168.1.1/49153 --> 192.168.1.11/9)
TX bitrate: 339588.28 kbit/s
RX bitrate: 340113.24 kbit/s
Mean Delay: 2.22 ms
Packet Loss Ratio: 0.00 %


./ns3 run scratch/pkt-11n-11ac-test -- 160 1 VhtMcs9
python3 scratch/flowmon-parse-results.py 80211acspeed.xml
FlowID: 1 (TCP 192.168.1.1/49153 --> 192.168.1.11/9)
TX bitrate: 364202.88 kbit/s
RX bitrate: 364748.85 kbit/s
Mean Delay: 2.04 ms
Packet Loss Ratio: 0.00 %
```

Oh, that's good! By configuring higher VHT MCS values at STAs, they enjoy HT results.

### Testing the MIMO and VHT MCS features together

Finally, let's test whether enabling both MIMO and higher MCS values can improve the throughput by executing the following command:

```
./ns3 run scratch/pkt-11n-test -- 160 2 VhtMcs9
python3 scratch/flowmon-parse-results.py 80211acspeed.xml
FlowID: 1 (TCP 192.168.1.1/49153 --> 192.168.1.11/9)
TX bitrate: 609000.27 kbit/s
RX bitrate: 609938.85 kbit/s
Mean Delay: 1.05 ms
Packet Loss Ratio: 0.00 %
```

Oh, that's good. Besides using channel bonding, we can see that enabling MIMO and adopting higher MCS values will improve the throughput for STAs significantly.

Having learned about ns-3 supporting 802.11n/ac features, let's start exploring 802.11ax and Wi-Fi 6 supporting features simulations using ns-3 without delay.

---

**802.11ac – channel bonding and FA**

The channel bonding and FA features are also supported in ns-3 802.11ac simulations. We just need to configure the Wi-Fi standard as 802.11ac, then we can test our simulation setup using channel widths of 80 MHz and 160 MHz and FA features. We have left this activity as an independent exercise for you.

---

# Exploring spectrum/resource management in Wi-Fi 6 using ns-3

The 802.11ax standard is also known as Wi-Fi 6 and high-efficiency Wi-Fi. The 802.11ax standard also supports all the advanced PHY and MAC features of 802.11n and ac, such as channel bonding, MIMO, higher MCS values, FA, and backward compatibility with past Wi-Fi standards, such as 802.11a, b, g, n, and ac. Equally, the ns-3 802.11ax implementation incorporated unique advanced features of the 802.11ax standard, such as resource scheduling, by adopting OFDMA technology at the PHY layer and resource scheduling at the MAC layer. In particular, resource scheduling in Wi-Fi networks is very helpful to support delay-sensitive smart applications. In this section, we will discuss how to simulate 802.11ax resource scheduling using the following hands-on activity:

- Setting up an 802.11ax-standard-configurable Wi-Fi LAN with 1 AP and 10 STAs

- Deploying the AP at location `100,100`

- Randomly deploying Wi-Fi STAs around the AP within a radius of 30 meters

- Configuring an 802.11ax-standard-specific spectrum and channels at a bandwidth of 40 MHz

- Configuring the 802.11ax MAC with a Round Robin scheduling algorithm for resource scheduling

- Configuring 10 STAs with TCP uplink flows using `OnOffApplication` at a constant bit rate of 10 Mbps

- Configuring and installing the `PacketSink` application at the AP to receive data from its STAs

- Testing the throughput and delay with no resource scheduling

- Testing the throughput and delay with resource scheduling using ns-3 and supporting 802.11ax's `RoundRobinScheduler`

## Setting up and simulating resource scheduling in 802.11ax/Wi-Fi 6 networks

In order to simulate 802.11ax Wi-Fi 6 networks, we use a similar topology to that used in `pkt-11n-11ac-test.cc`.

Let's start the 802.11ax network setup in `pkt-11ax-sched.cc` by importing the following necessary packages for spectrum modeling, besides all the other Wi-Fi-related packages in `pkt-11n-11ac-test.cc`:

```
#include "ns3/spectrum-wifi-helper.h"
#include "ns3/multi-model-spectrum-channel.h"
#include "ns3/wifi-acknowledgment.h"
```

In `main()`, copy the Wi-Fi topology setup-related line of code from `pkt-11n-11ac-test.cc` into `pkt-11ax-sched.cc`.

Then, configure our simulation topology Wi-Fi standard as the 802.11ax standard and its specific spectrum, and create a spectrum channel using the following lines of code:

```
WifiHelper wifi;
wifi.SetStandard (WIFI_STANDARD_80211ax);
Ptr<MultiModelSpectrumChannel> spectrumChannel =
CreateObject<MultiModelSpectrumChannel> ();
SpectrumWifiPhyHelper wifiPhy;
wifiPhy.SetPcapDataLinkType (WifiPhyHelper::DLT_IEEE802_11_
RADIO);
wifiPhy.SetChannel (spectrumChannel);
```

Next, set the operating channel using the following lines of code:

```
wifiPhy.Set ("ChannelSettings", StringValue ("{62,
40,   BAND_5GHZ, 0}"));
```

After completing the 802.11ax-specific PHY layer configuration, start the 802.11ax-specific MAC configuration by setting NS3's `RrMultiUserScheduler` as a resource scheduling algorithm using the following lines of code:

```
WifiMacHelper wifiMac;
wifiMac.SetMultiUserScheduler ("ns3::RrMultiUserScheduler",
                    "EnableUlOfdma", BooleanValue
(true),
                    "EnableBsrp", BooleanValue (true));
```

Then, install the 802.11ax-specific PHY, MAC, and `InternetStack` protocols, and configure the IP addresses for all Wi-Fi nodes (the AP and STAs) (this is similar to `pkt-11n-11ac-test.cc`'s code).

With that, we have completed our 802.11ax network setup. Let's start installing suitable applications for testing it.

First, install the TCP-socket-based `PacketSink` application on the AP node to receive TCP data from Wi-Fi STAs using the following lines of code:

```
uint64_t port = 9;
Address localAddress (InetSocketAddress (Ipv4Address::GetAny
(), port));
```

```
    PacketSinkHelper sink ("ns3::TcpSocketFactory",
localAddress);
    ApplicationContainer apps_sink = sink.Install (APs.Get (0));
```

Next, install a TCP OnOffApplication on all STAs to send data at a constant rate of 10 Mbps to the AP using the following lines of code:

```
    for (int i=0;i<nStas;i++)
    {
      OnOffHelper onoff ("ns3::TcpSocketFactory",
InetSocketAddress (apInterface.GetAddress(0), port));
      onoff.SetConstantRate (DataRate ("10Mbps"), 1000);
      onoff.SetAttribute ("StartTime", TimeValue (Seconds (6)));
      onoff.SetAttribute ("OnTime",  StringValue
("ns3::ConstantRandomVariable[Constant=1]"));
      onoff.SetAttribute ("OffTime", StringValue
("ns3::ConstantRandomVariable[Constant=0]"));
      onoff.SetAttribute ("StopTime", TimeValue (Seconds (8)));
      ApplicationContainer apps_source = onoff.Install (stas.Get
(i));
      apps_sink.Start (Seconds (0.5));
      apps_sink.Stop (Seconds (10));
    }
```

After writing the testing code, close main() by freeing all the simulation resources, and configure a flow monitor to analyze the simulation results, similar to the pkt-11n-11ac-test.cc code snippets.

That's all! We have successfully configured an 802.11ax simulation setup in ns-3 and are ready to test it. Let's evaluate 802.11ax scheduling in the next section.

## Evaluating the resource scheduling feature of 802.11ax

We will start our 802.11ax simulation setup evaluation without any radio resource schedulers.

### Testing the simulation setup without the resource scheduling feature

Let's test our 802.11ax simulation setup without a resource scheduling algorithm (comment the resource scheduling configuration in pkt-11ax-sched.cc) using the following command in the ns-3 installation directory:

```
./ns3 run scratch/pkt-11ax-sched
python3 scratch/flowmon-parse-results.py wlan11axspeed.xml >
noscheduling
```

We collect the flow statistics into a noscheduling file. Then, we can get the average throughput using the following commands:

```
cat noscheduling |grep 'RX bitrate' | cut -f3 -d ' ' | head -n
10| awk '{sum=sum+$1}END{print sum/10}'
7.5Mbps
```

We collect the flow statistics into the noscheduling file. Then, we can get the average delay using the following commands:

```
cat noscheduling |grep 'Mean Delay' | cut -f3 -d ' ' | head -n
10| awk '{sum=sum+$1}END{print sum/10}'
27.8 ms
```

From the results, we observe that the STAs' average throughput is 7.5 Mbps and they experience an average delay of 27.8 ms.

### Testing with resource scheduling

Let's enable resource scheduling by uncommenting the resource scheduler configuration code in pkt-11ax-sched.cc and running the following commands:

```
./ns3 run scratch/pkt-11ax-sched
python3 scratch/flowmon-parse-results.py wlan11axspeed.xml >
scheduling
```

We collect the flow statistics into the scheduling file. Then, we can get the average throughput using the following commands:

```
cat scheduling |grep 'RX bitrate' | cut -f3 -d ' ' | head -n
10| awk '{sum=sum+$1}END{print sum/10}'
8.5Mbps
```

We collect flow statistics into the scheduling file. Then, we can get the average delay using the following commands:

```
cat scheduling |grep 'Mean Delay' | cut -f3 -d ' ' | head -n
10| awk '{sum=sum+$1}END{print sum/10}'
11.48ms
```

Oh, that's great. We can clearly observe that enabling resource scheduling significantly decreased the average delay for STAs to 11.48 ms from 27.8 ms. Resource scheduling also helps STAs experience an average throughput of 8.5 Mbps, which is 1 Mbps higher than without resource scheduling.

Congratulations. We have successfully set up an 802.11ax network and evaluated the resource-scheduling performance in our simulation setup. Next, we will discuss another interesting ns-3-supported 802.11ax feature: BSS coloring.

## Exploring spectrum reuse in Wi-Fi 6 using ns-3

In various hotspots, such as airports, shopping malls, and stadiums, it is necessary to have dense Wi-Fi 6 network deployments. Operators may go for a spectrum reuse deployment to maximize channel utilization. However, the dense deployment of Wi-Fi networks with frequency-reuse 1 could lead to high interference in the network. Hence, usually, we deploy Wi-Fi networks with careful channel planning. Still, parallel transmissions from neighboring Wi-Fi networks could lead to exposed terminals and unnecessary channel access delays. This affects the networks' throughput significantly. In 802.11ax wireless networks, these issues can be limited by enabling careful BSS coloring, such as cellular network base stations operating channel-frequency planning. ns-3 for 802.11ax can also simulate the BSS coloring feature.

In order to analyze the BSS coloring feature of Wi-Fi 6, we planned a hands-on activity with the following activities:

- Setting up two BSSs – BSS-1 (SSID: `Packt-1`) and BSS-2 (SSID: `Packt-2`)
- Deploying BSS-1 with 1 AP at location (`100,100`) and its STAs around the AP within a radius of 20 meters
- Deploying BSS-2 with 1 AP at location (`100,150`) and its STAs around the AP within a radius of 20 meters
- Configuring the 802.11ax standard and the same operating channel for both BSSes
- Configuring an **Overlapping BSS/Preamble Detection** (**OBSS/PD**) algorithm to increase parallel transmissions
- Configuring important parameters for the OBSS/PD algorithm to work, such as the OBSSPD threshold, minimum RSSI threshold, and CCA threshold, and the Tx power and Rx sensitivity of APs and STAs at their specific PHY layers
- Configuring UDP-socket-based `OnOffApplication` parallel traffic flows at BSS-1 and BSS-2 to evaluate how BSS coloring and the OBSS/PD algorithm improved the performance of 802.11ax networks
- Testing the throughput for STAs when disabling BSS coloring and the OBSS/PD algorithm and the number of STAs affected due to parallel transmissions
- Testing the throughput for STAs when enabling BSS coloring and the OBSS/PD algorithm and the number of STAs affected due to parallel transmissions

## Setting up and simulating resource scheduling in 802.11ax/Wi-Fi6 networks

Let's set up an 802.11ax Wi-Fi LAN (`pkt-11ax-bss-color.cc`) with two BSSes to evaluate the performance with BSS coloring and the OBSS/PD algorithm in ns-3.

1. Let's import all the following necessary packages for setting up the 802.11ax network and carrying out the hands-on activities:

```
#include "ns3/core-module.h"
#include "ns3/point-to-point-module.h"
#include "ns3/network-module.h"
#include "ns3/applications-module.h"
#include "ns3/internet-apps-module.h"
#include "ns3/mobility-module.h"
#include "ns3/csma-module.h"
#include "ns3/internet-module.h"
#include "ns3/yans-wifi-helper.h"
#include "ns3/ssid.h"
#include "ns3/spectrum-wifi-helper.h"
#include "ns3/multi-model-spectrum-channel.h"
#include "ns3/wifi-net-device.h"
#include "ns3/ap-wifi-mac.h"
#include "ns3/he-configuration.h"
#include "ns3/flow-monitor-module.h"
using namespace ns3;
```

2. Start `main()` by creating the necessary number of APs and STAs for setting up the two 802.11ax-standard-based BSSes (BSS-1 and BSS-2):

```
int
main (int argc, char *argv[])
{
 uint32_t nWifi = 20;
 NodeContainer WiFiAPNodes;
 WiFiAPNodes.Create (2);
 NodeContainer wifiStaNodes1;
 wifiStaNodes1.Create (nWifi);
 NodeContainer wifiStaNodes2;
 wifiStaNodes2.Create (nWifi);
```

3.  Then, set up the BSS-1 topology as per our hands-on activity using the following lines of ns-3 code:

```
NodeContainer wifiApNode1 = WiFiAPNodes.Get (0);
MobilityHelper mobility;
Ptr<ListPositionAllocator> positionAlloc =
CreateObject<ListPositionAllocator> ();
positionAlloc->Add (Vector (100, 100, 0.0));
mobility.SetPositionAllocator (positionAlloc);
mobility.SetMobilityModel
("ns3::ConstantPositionMobilityModel");
mobility.Install (wifiApNode1);
mobility.SetPositionAllocator
("ns3::RandomDiscPositionAllocator",
                            "X", StringValue ("100.0"),
                            "Y", StringValue ("100.0"),
                            "Rho", StringValue
("ns3::UniformRandomVariable[Min=0|Max=20]"));
mobility.Install (wifiStaNodes1);
```

4.  Similarly, set up the BSS-2 topology as per our hands-on activity using the following lines of ns-3 code:

```
NodeContainer wifiApNode2 = WiFiAPNodes.Get (1);
MobilityHelper mobility1;
Ptr<ListPositionAllocator> positionAlloc1 =
CreateObject<ListPositionAllocator> ();
positionAlloc1->Add (Vector (100, 150, 0.0));
mobility1.SetPositionAllocator (positionAlloc1);
mobility1.SetMobilityModel
("ns3::ConstantPositionMobilityModel");
mobility1.Install (wifiApNode2);
mobility1.SetPositionAllocator
("ns3::RandomDiscPositionAllocator",
                            "X", StringValue ("100.0"),
                            "Y", StringValue ("150.0"),
                            "Rho", StringValue
("ns3::UniformRandomVariable[Min=0|Max=20]"));
mobility1.Install (wifiStaNodes2);
```

5. Next, connect both BSSes using the P2P channel using the following lines of code:

```
PointToPointHelper pointToPoint;
pointToPoint.SetDeviceAttribute ("DataRate",
StringValue ("5Mbps"));
pointToPoint.SetChannelAttribute ("Delay", StringValue
("2ms"));
NetDeviceContainer p2pDevices;
p2pDevices = pointToPoint.Install (WiFiAPNodes);
```

That's good. We have set up the BSS-1 and BSS-2 topologies. Next, let's start configuring the PHY and MAC layers for both BSSes sequentially:

1. Let's start by setting the 802.11ax Wi-Fi standard and its specific spectrum model for both BSSes using the following ns-3 lines of code. This also involves creating a common channel and setting channel details such as the operating frequency, channel width, and channel number:

```
WifiHelper wifi;
Wifi.SetStandard (WIFI_STANDARD_80211ax);
Ptr<MultiModelSpectrumChannel> spectrumChannel =
CreateObject<MultiModelSpectrumChannel> ();
SpectrumWifiPhyHelper wifiPhy;
wifiPhy.SetPcapDataLinkType (WifiPhyHelper::DLT_
IEEE802_11_RADIO);
wifiPhy.SetChannel (spectrumChannel);
wifiPhy.Set ("ChannelSettings", StringValue ("{42, 80,
BAND_5GHZ, 0}"));
```

2. Next, set the OBSS/PD algorithm to enable dynamic preamble detection (SD detection) to improve the number of parallel transmissions:

```
wifi.SetObssPdAlgorithm ("ns3::ConstantObssPdAlgorithm",
                         "ObssPdLevel", DoubleValue
(obssPdThreshold));
```

3. Now, configure the STA-specific PHY parameters for BSS-1, such as Tx power, Rx sensitivity, and CCA threshold, using the following lines of code:

```
wifiPhy.Set ("TxPowerStart", DoubleValue (powSta));
wifiPhy.Set ("TxPowerEnd", DoubleValue (powSta));
wifiPhy.Set ("CcaEdThreshold", DoubleValue
(ccaEdTrSta));
wifiPhy.Set ("RxSensitivity", DoubleValue (-92.0));
```

4.  Next, configure the STA-specific MAC for BSS-1 with the SSID (`Packt-1`) and install the PHY and MAC layers using the following lines of code:

```
WifiMacHelper mac;
Ssid ssid1 = Ssid ("Packt-1");
NetDeviceContainer staDevices1;
mac.SetType ("ns3::StaWifiMac",
             "Ssid", SsidValue (ssid1),
             "ActiveProbing", BooleanValue (false));
staDevices1 = wifi.Install (wifiPhy, mac,
wifiStaNodes1);
```

5.  After configuring the STA Wi-Fi protocol for BSS-1, configure the AP-specific PHY parameters for it, such as Tx power, Rx sensitivity, and the CCA thresholds, using the following lines of code:

```
wifiPhy.Set ("TxPowerStart", DoubleValue (powAp));
wifiPhy.Set ("TxPowerEnd", DoubleValue (powAp));
wifiPhy.Set ("CcaEdThreshold", DoubleValue
(ccaEdTrAp));
wifiPhy.Set ("RxSensitivity", DoubleValue (-92.0));
NetDeviceContainer apDevices1;
```

6.  Configure the AP-specific MAC with the SSID (`Packt-1`) and install the PHY and MAC layers using the following lines of code:

```
mac.SetType ("ns3::ApWifiMac",
             "Ssid", SsidValue (ssid1));
apDevices1 = wifi.Install (wifiPhy, mac, wifiApNode1);
```

7.  Now, we can set the AP-specific BSS color to (1) using the following lines of code:

```
Ptr<WifiNetDevice> apDevice = apDevices1.Get
(0)->GetObject<WifiNetDevice> ();
Ptr<ApWifiMac> apWifiMac = apDevice->GetMac
()->GetObject<ApWifiMac> ();
apDevice->GetHeConfiguration ()->SetAttribute
("BssColor", UintegerValue (1));
```

8.  After setting up the BSS-1 PHY and MAC layers and BSS-related configuration, similarly, configure and install the 802.11ax-specific PHY and MAC layers and BSS coloring for the BSS-2 STAs and APs using the following lines of code in ns-3:

```
Ssid ssid2 = Ssid ("Packt-2");
NetDeviceContainer staDevices2;
mac.SetType ("ns3::StaWifiMac",
            "Ssid", SsidValue (ssid2),
            "ActiveProbing", BooleanValue (false));
staDevices2 = wifi.Install (wifiPhy, mac,
wifiStaNodes2);
wifiPhy.Set ("TxPowerStart", DoubleValue (powAp));
wifiPhy.Set ("TxPowerEnd", DoubleValue (powAp));
wifiPhy.Set ("CcaEdThreshold", DoubleValue
(ccaEdTrAp));
wifiPhy.Set ("RxSensitivity", DoubleValue (-92.0));
NetDeviceContainer apDevices2;
mac.SetType ("ns3::ApWifiMac",
            "Ssid", SsidValue (ssid2));
apDevices2 = wifi.Install (wifiPhy, mac, wifiApNode2);
Ptr<WifiNetDevice> apDevice2 = apDevices2.Get
(0)->GetObject<WifiNetDevice> ();
Ptr<ApWifiMac> apWifiMac2 = apDevice2->GetMac
()->GetObject<ApWifiMac> ();
apDevice2->GetHeConfiguration ()->SetAttribute
("BssColor", UintegerValue (2));
mobility.Install (wifiAPNodes);
```

That's all we need to enable the BSS coloring feature and deploy the 802.11ax network with frequency-reuse 1.

9.  Next, before installing the testing applications, install `InternetStack` and configure the IP addresses for all APs and STAs using the following lines of code:

```
Ipv4AddressHelper address;
address.SetBase ("10.1.1.0", "255.255.255.0");
Ipv4InterfaceContainer p2pInterfaces;
Ipv4InterfaceContainer apInterfaces1;
Ipv4InterfaceContainer apInterfaces2;
Ipv4InterfaceContainer staInterfaces1;
```

```
Ipv4InterfaceContainer staInterfaces2;
p2pInterfaces = address.Assign (p2pDevices);
address.SetBase ("10.1.3.0", "255.255.255.0");
apInterfaces1=address.Assign (apDevices1);
staInterfaces1=address.Assign (staDevices1);
//address.SetBase ("10.1.4.0", "255.255.255.0");
apInterfaces2=address.Assign (apDevices2);
staInterfaces2=address.Assign (staDevices2);
```

10. Let's start testing our simulation topology to evaluate BSS coloring and its benefits. For testing, we configure the UDP-based OnOffApplications on STAs to send traffic at a constant data rate to their APs using the following lines of code. At the APs, PacketSink applications are configured to receive data using the following lines of code:

```
uint8_t port1=9;
Address localAddress (InetSocketAddress
(Ipv4Address::GetAny (), port1));
PacketSinkHelper sink ("ns3::UdpSocketFactory",
localAddress);
ApplicationContainer apps_sink = sink.Install
(wifiApNode1);
apps_sink.Start (Seconds (6.0));
apps_sink.Stop (Seconds (10.0));
for (int i=0;i<20;i++)
{
   OnOffHelper onoff ("ns3::UdpSocketFactory",
InetSocketAddress (apInterfaces1.GetAddress(0), port1));
   onoff.SetConstantRate (DataRate ("10Mbps"), 1000);
   onoff.SetAttribute ("StartTime", TimeValue (Seconds
(6)));
   onoff.SetAttribute ("StopTime", TimeValue (Seconds
(8)));
   onoff.SetAttribute ("OnTime",  StringValue
("ns3::ConstantRandomVariable[Constant=1]"));
   onoff.SetAttribute ("OffTime", StringValue
("ns3::ConstantRandomVariable[Constant=0]"));
   ApplicationContainer apps_source = onoff.Install
(wifiStaNodes1.Get (i));
}
```

```
int8_t port2=10;
Address localAddress2 (InetSocketAddress
(Ipv4Address::GetAny (), port2));
PacketSinkHelper sink2 ("ns3::UdpSocketFactory",
localAddress2);
ApplicationContainer apps_sink2 = sink2.Install
(wifiApNode2);
apps_sink2.Start (Seconds (6.0));
apps_sink2.Stop (Seconds (10.0));
for(int i=0;i<20;i++)
{
   OnOffHelper onoff2 ("ns3::UdpSocketFactory",
InetSocketAddress   (apInterfaces2.GetAddress(0),
port2));
onoff2.SetConstantRate (DataRate ("10Mbps"), 1000);
onoff2.SetAttribute ("StartTime", TimeValue (Seconds
(6)));
onoff2.SetAttribute ("StopTime", TimeValue (Seconds
(8)));
onoff2.SetAttribute ("OnTime",  StringValue
("ns3::ConstantRandomVariable[Constant=1]"));
onoff2.SetAttribute ("OffTime", StringValue
("ns3::ConstantRandomVariable[Constant=0]"));
ApplicationContainer apps_source = onoff2.Install
(wifiStaNodes2.Get (i));
}
```

Here, we need to observe both BSSes starting their OnOffApplication objects: onoff and onoff2 at the same time. Hence, we can simulate parallel transmission in the network.

11. Finally, close main() by freeing all the simulation resources, and configure a flow monitor to analyze the simulation results:

```
Ptr<FlowMonitor> flowmon;
FlowMonitorHelper flowmonHelper;
flowmon = flowmonHelper.InstallAll ();
Simulator::Stop (Seconds (15.0));
Simulator::Run ();
flowmon->SerializeToXmlFile ("11axbsscolor.xml", true,
true);
```

```
        Simulator::Destroy ();
        return 0;
}
```

That's all – we have successfully configured an 802.11ax simulation setup for testing the BSS coloring feature in ns-3.

## Evaluation of the BSS coloring feature

Before evaluating the 802.11ax BSS coloring feature, let's evaluate our simulation setup without the BSS coloring feature.

### Testing without BSS coloring

Let's start testing our simulation setup by disabling BSS coloring features (comment all BSS coloring and OBSS/PD algorithm-related lines of code for the PHY and MAC layers). Run the following commands in your ns-3 installation directory:

```
./ns3 run scratch/pkt-11ax-bss-color
python3 scratch/flowmon-parse-results.py 11axbsscolor.xml >
bsscolorfalse
cat bsscolorfalse |grep 'RX' | cut -f3 -d ' ' | awk '{if
($1>10000) print $1;}'
None
14449.19
21990.75
154200.00
None
...
```

> **Incorrect Rx rates at APs observed!**
>
> In our simulation setup, the STAs' `OnOffApplication` objects `onoff` and `onoff2` are sending a max data rate of 10 Mbps only. However, at the AP, we see higher Rx rate values such as 14 Mbps, 21 Mbps, and **None**. Higher Rx rate values are due to parallel transmissions intermixing at the AP. Hence, we should consider these Rx rate values incorrect and ignore these values. Besides, some STAs' application traffic flows are not at all receiving at AP. This means all these STAs are affected due to the frequency-reuse dense deployment and parallel transmissions.

We collect the flow stats into the `bsscolorfalse` file. Let's find how many total STAs are affected when BSS coloring is disabled:

```
cat bsscolorfalse |grep 'RX' | cut -f3 -d ' ' | awk '{if
($1>10000) print $1;}' |wc -1
15
```

For a total of 15 STAs out of 40 STAs, parallel transmissions are affected in a frequency-reuse 1 dense deployment.

### Testing our setup by enabling BSS coloring

Now, let's start testing our simulation setup by enabling BSS coloring features using the following commands and collecting flow stats into the `bsscolortrue` file:

```
./ns3 run scratch/pkt-11ax-bss-color
python3 scratch/flowmon-parse-results.py 11axbsscolor.xml >
bsscolortrue
cat bsscolortrue |grep 'RX' | cut -f3 -d ' ' | awk '{if
($1>10000) print $1;}' | wc -1
6
```

If the BSS coloring feature is enabled, parallel transmissions are affected in frequency-reuse 1 dense deployment for only 6 out of 40 STAs.

That's superb. The 802.11ax BSS coloring feature is highly useful for maximizing parallel transmissions with a frequency-reuse 1 dense deployment.

## Summary

Well done. We successfully explored advanced Wi-Fi standards such as 802.11n, ac, and ax and their features using ns-3 simulations. ns-3 helps readers explore and research how to improve the throughput for Wi-Fi STAs and their response times using advanced Wi-Fi PHY and MAC features such as channel bonding, FA, MIMO, and higher MCS values.

In particular, by learning about 802.11ax features such as resource scheduling and BSS coloring simulations, readers can carry out their research activities systematically to set up and evaluate real-time, dense Wi-Fi deployments, such as airports, stadiums, shopping malls, and IoT use cases. Overall, this chapter mainly helped readers approach various advanced Wi-Fi features, such as scheduling, interference management, and channel utilization improvement techniques.

In the next chapter, we will discuss another interesting advanced network simulation topic – 4G/LTE network simulations.

# Part 3: Learn, Set Up, and Evaluate 4G Long-Term Evolution (LTE) Networks

On completion of *Part 3*, you will have learned how to install, configure, and test important network nodes and protocols of LTE radio access and core networks using ns-3. As part of LTE research activities, you will learn how to set up and simulate LTE radio resource management, mobility scenarios, and interference handling using ns-3-supporting LTE algorithms.

This part has the following chapters:

- *Chapter 7, Getting Started with LTE Network Simulations Using ns-3*
- *Chapter 8, Researching LTE Network Radio Resource Management and Mobility Management Using ns-3*
- *Chapter 9, Researching LTE Advanced Networks: LTE HetNets and Interference Management Using ns-3*

# 7

# Getting Started with LTE Network Simulations Using ns-3

In order to cope with the explosion of data traffic demands from a huge number of mobile subscribers, 3GPP came up with next-generation (4G) **long-term evolution** (**LTE**) mobile networks. Unlike 2G/3G traditional circuit-switched systems and packet data networks, LTE aims at offering both voice and data services using packet-switched networks. Broadly, LTE architecture comprises the core network (**Evolved Packet Core** (**EPC**)) and the **radio access network** (**RAN**). Mainly, LTE supports the following major features:

- Higher data rates (downlink transmission speed around 100 Mbps and uplink transmission speed around 50 Mbps)

- Lower connection establishment delay (<100 ms) and transmission latencies (<10 ms)

- LTE mobile networks can be deployed in the following range of operating bandwidths: 1.4 to 20 MHz

- Simplified network architecture to easily upgrade necessary data plane or control plane networking equipment

- Supports seamless mobility for users, including different radio-access technologies such as 2G, 3G, and non-3GPP technologies

- Supports **multiple input, multiple output** (**MIMO**) to improve spectral efficiency (in uplink MIMO, up to 7.5 bps/Hz, and in downlink MIMO, up to 30 bps/Hz)

- Promising to improve cell-edge users' bit rates for ensuring fairness in the network

- Improving the energy efficiency of mobile terminals/devices

All these unprecedented features motivate many operators to transition from their existing mobile networks to LTE or upgrade to LTE network deployments. However, deploying LTE networks involves challenges for operators in terms of capital and operational expenditures. Hence, it is important to systematically study and evaluate various LTE network deployment scenarios before going for real-time deployments. In order to support 4G network researchers, engineers, and operators, in this chapter, we introduce important features of LTE networks supported in ns-3 and how to simulate LTE networks. We start with discussing how ns-3 supports LTE network architecture and its important components. Specifically, we discuss ns-3 supporting LTE EPC and RAN components, their protocols, and interfaces to set up a variety of LTE networks. Then, we discuss how to create and configure various LTE network components using ns-3 code snippets. Finally, we conclude this chapter with a hands-on activity as a step-by-step procedure to set up complete LTE EPS comprising EPC and RAN topology and evaluate it using ns-3-supporting network applications. Mainly, we discuss the following topics in this chapter:

- Understanding the LTE network nodes, protocols, and features supported in ns-3
- Learning how to set up and configure various LTE nodes in ns-3
- Simulating LTE RAN and EPC networks using ns-3 key building blocks

## Understanding the LTE network nodes, protocols, and features supported in ns-3

ns-3 LTE module implementation considers the following broad architecture for simulations. We can understand it in two major parts, as shown in *Figure 7.1*:

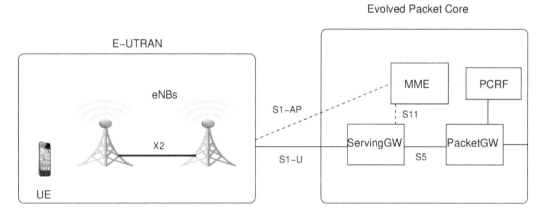

Figure 7.1 – ns-3 LTE architecture implementation supporting nodes and their interfaces

The first part is the ns-3 implementation of the **Evolved Universal Terrestrial Radio Access Network (E-UTRAN)** and its functions. It mainly deals with LTE spectrum management, LTE base stations, **evolved Node B (eNB)** and its protocols, LTE mobile terminals, and **user equipment (UE)** and its protocols. The second part is the ns-3 implementation of the LTE EPC network. It mainly deals with EPC's major network components such as **Mobility Management Entity (MME)**, the **Packet Gateway (P-GW)**, and **Serving Gateway (S-GW)**, and their protocols.

ns-3 LTE implementation considers all major interfaces such as S1-U, S5, S1-AP, S11, and X2 to connect E-UTRAN and EPC network components. Together, LTE RAN and EPC are known as LTE **Evolved Packet System (EPS)**. In this section, we start with discussing the ns-3 LTE module supporting E-UTRAN components and their protocols for setting up LTE networks and conducting simulations. Please note that in this chapter, we assume users have basic knowledge of LTE networks. Hence, to quickly introduce ns-3 LTE simulations, we focus on introducing high-level LTE details and supporting features of each of the ns-3 LTE components and their protocols. In the upcoming section, we start by discussing ns-3 supporting the E-UTRAN protocol stack, and eNB and UE implementation details.

## ns-3 LTE implementation supporting E-UTRAN features

LTE E-UTRAN is known as an access network, which contains a number of eNBs and pieces of UE to access LTE services for browsing the internet or making voice calls. In order to simulate eNB and UE features and operations, ns-3 LTE module implementation focuses on the following LTE standard protocol layers.

The LTE-specific protocol stack contains the following layers:

- **Radio Resource Control (RRC)**
- **Packet Data Convergence Protocol (PDCP)**
- **Radio-Link Control (RLC)**
- **Medium Access Control (MAC)**
- LTE **Physical layer (PHY)** (also known as **L1**)

In the LTE network, all data transmissions are processed through the data plane radio protocols stack (PDCP, RLC, MAC, and L1) of eNBs and UE. And all control signals related to connection management, radio resources management, **quality of service (QoS)** bearers' management for handling voice, video, and a variety of data traffic flows, and mobility management are processed through the control plane radio protocols stack (RRC, PDCP, RLC, MAC, and L1) of eNB and UE. That means either LTE data or control signals must process through the PDCP, RLC, MAC, and L1 stack. The ns-3 LTE module also supports LTE data plane and control plane protocol stacks with few limitations. Let's quickly go through ns-3 supporting features of the LTE data and control plane stacks.

### ns-3 LTE data plane protocol stack and its supporting features

Let's discuss each of the ns-3 supporting LTE protocols one by one from top to bottom in the protocol stack order:

- **ns-3 PDCP layer**: In LTE RAN, every application traffic flow (IP packet) is mapped to a specific radio bearer to ensure their QoS requirements in terms of bit rate and latencies. The ns-3 LTE data plane stack receives IP packets from the TCP/IP layer to the PDCP layer. In ns-3, PDCP layer implementation is simplified. It maintains the **sequence number (SN)** of PDCP packets during user plane or control plane data transmissions. It also supports the handover procedure by transferring the SN status between the source eNB and destination eNB. In ns-3 LTE simulations, as per standard, one PDCP entity is mapped to one radio bearer for processing its IP packets.

  However, the following LTE PDCP standard features are not supported in the current ns-3.36 LTE module:

  - Compression and decompression of IP data flows

  - Encryption and decryption of user plane data and control plane data

  - Data integrity

  - Duplicate packets handling

- **ns-3 RLC layer**: Similar to PDCP, the ns-3 LTE RLC entity is also mapped to one radio bearer configured for UE. Mainly, RLC provides services to the PDCP layer at the radio bearer level. It performs segmentation and concatenation of transmitting data by coordinating with the MAC layer scheduler for getting possible transmission sizes. RLC also ensures retransmissions, duplicate detections, and in-order delivery to higher layers. Mainly, the ns-3 RLC layer is implemented as per 3GPP technical specifications and it supports the following modes in simulations: **Transparent Mode (TM)**, **Unacknowledged Mode (UM)**, and **Acknowledged Mode (AM)**.

  However, the following LTE RLC layer standard features are not implemented in the current ns-3.36 LTE module:

  - The **Saturation Mode** RLC is not implemented in the case of LTE RAN and EPC simulations

  - No notification from RLC to PDCP layers in case of success or failure in AM transmissions

- **ns-3 MAC layer**: ns-3 LTE MAC mainly deals with radio resources scheduling and random access procedure for UE connections. The ns-3 LTE module supports a variety of radio resource schedulers for **uplink (UL)** (UL: UE to eNB) and **downlink (DL)** (DL: eNB to UE) transmissions. In order to inform UE about resource block allocation on a per subframe or **transmission time interval (TTI)** (1 ms) basis, ns-3 schedulers generate **data control indication (DCI)** structures and then transmit them to the PHY layer of the eNB. ns-3 LTE MAC also supports random access procedures for UE.

However, it is simplified and handled with the following messages only:

- **Random Access (RA)** request: This is simulated by sending an RA preamble using an ideal message (ns-3 LTE control message), which does not need any radio resources. Moreover, it uses RA channel configuration index 14. Hence, RA preambles can be sent on any system frame number and subframe number.

- RA response: This is simulated by sending an ideal message (ns-3 LTE control message), which does not need any radio resources.

- LTE message 3: This is modeled as a real RLC TM **Protocol Data Unit** (**PDU**) whose UL resources are allocated by the ns-3 scheduler.

- RA preambles' contention resolution is not modeled.

- **ns-3 LTE PHY layer**: This mainly handles various DL and UL channels, which are major interfaces between eNBs and UE:

  - ns-3 LTE PHY mainly simulates the following downlink channels: **Physical Downlink Shared Channel** (**PDSCH**) for unicast data transmission, **Physical Downlink Control Channel** (**PDCCH**) for sending downlink control information and sharing scheduling details, and **Physical Control Format Indicator Channel** (**PCFICH**) for providing the UE with the necessary information to decode the set of PDCCHs.

  - This layer simulates the **Physical Uplink Shared Channel** (**PUSCH**), which is similar to the PDSCH for handling UL transmissions of UE. There is at most one PUSCH per uplink component carrier per UE. ns-3 also supports the **Physical Uplink Control Channel** (**PUCCH**) for sending hybrid-ARQ acknowledgments from the UE to eNBs to indicate whether the downlink data is successfully received or not.

  - It also simulates the **Physical Random-Access Channel** (**PRACH**) for supporting RA procedures for UE.

## ns-3 LTE control plane protocol stack and its supporting features

LTE RRC layers at eNBs and UE together handle all radio resource control handling activities, such as sharing broadcasting cell configuration details, handling connections, handover-related measurement configurations, and radio bearers' configurations. Let's discuss ns-3 LTE RRC supporting features.

RRC is considered a radio resource control signaling handling protocol. It handles the UEs' connection setup and mobility functions. LTE control messages are transmitted to the UEs from either MME or eNBs.

The ns-3 RRC handles the following RAN-related procedures:

- Generation of **Master Information Block** (**MIB**), **System Information Block** (**SIB**)-1, and SIB-2 messages at eNBs. MIB contains configuration parameters related to the PHY layer generated during cell configuration, and it is broadcast every 10 ms at the beginning of the radio frame as a control message. SIB-1 provides information necessary for network access cell selection evaluation and it is broadcast every 20 ms. SIB-2 provides uplink and **Random Access Channel** (**RACH**)-related configurations.

- Interpretation of MIB, SIB-1, and SIB-2 messages at UE.

- Initial cell selection at UE.

- Handling the UE RRC connection establishment procedure.

- RRC reconfiguration procedures for handling handovers, UE handover measurement configurations, and radio bearers reconfiguration.

- RRC connection reestablishment for handling handover procedures.

The ns-3.36 RRC has the following limitations:

- Not offering security for control signaling transmissions

- Assumes no transmission errors for control signaling

- Multiple carrier frequencies are not supported

- Simulating multiple network operators is not supported

- Simulating marking a cell as barred or reserved is not supported

Note that any control signals generated by RRC are also processed through the data plane protocol stack (PDCP, RLC, MAC, and L1) of UE and eNBs.

Next, let's discuss ns-3 LTE eNB supporting operations and functionalities.

## ns-3 LTE eNB operations and functionalities

In ns-3 LTE simulations, one of the most important components is eNB. In ns-3, eNBs can be easily simulated by using ns-3 nodes with LTE data plane and control plane protocols. Here, eNBs' data plane protocols (refer to *Figure 7.2*) handle tasks related to IP packet exchange between UE and EPC through S-GW. ns-3 eNB connects with the ns-3 S-GW node using the S1-U interface and their data exchange is done through the GTP-U, UDP/IP, L2, and L1 protocol stack. Hence, ns-3 eNBs are able to transmit UE traffic flow to the EPC by establishing necessary tunnels between UE and eNB, and eNB and S-GWs. ns-3 eNBs use the PDCP, RLC, MAC, and L1 protocol stack while exchanging data traffic between UE and eNBs.

On other hand, ns-3 eNBs' control plane protocols (refer to *Figure 7.2*) handle all control messages between UE and eNBs, and eNBs and MME. In order to exchange all radio resource-related control signals, relevant messages are exchanged between UE and eNBs through the RRC, PDCP, RLC, MAC, and L1 protocol stack. Similar to the LTE standard, ns-3 eNBs are connected to the ns-3 MME node using the S1-AP interface to exchange necessary control messages between eNBs and MME through the S1-AP, SCTP, IP, L2, and L1 protocol stack. Besides, ns-3 eNBs can be connected to each other using an X2 (logical link) interface for communication between eNBs to handle mobility and interference management activities.

Let's see the ns-3 LTE module implementation of eNB's data plane and control plane in *Figure 7.2*:

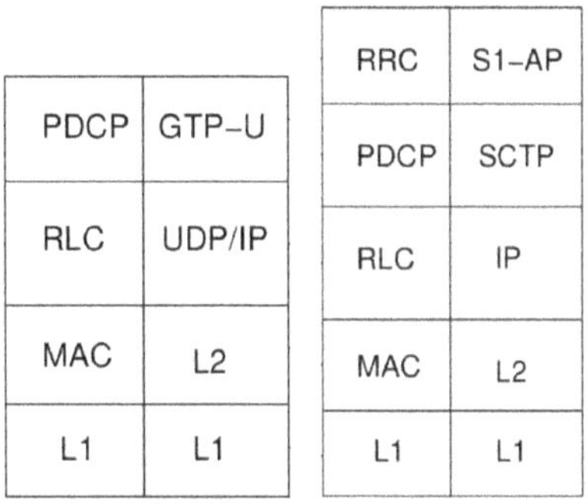

Figure 7.2 – ns-3 eNB data plane and control plane protocols

> **Important note**
> In the LTE standard, one eNB can be connected to multiple MMEs/S-GWs for the purpose of load sharing and redundancy. However, in ns-3 LTE simulations, one eNB can be connected to only one MME and one S-GW.

ns-3 eNBs handle the following major operations matching with the LTE standard:

- Operates in FDD mode at flexible bandwidths ranging from 6 **resource blocks** (**RBs**) to 100 RBs

- Broadcasts MIB and SIBs at specific intervals

- Radio admission control: For UE RRC connection requests, the eNB RRC layer replies with SETUP (success) or REJECT messages for assigning suitable RBs to UE by a variety of schedulers

- Supports both uplink and downlink schedulers

- Radio bearers control radio bearer configuration. ns-3 eNB supports the following three **logical channel groups** (**LCGs**) for handling radio bearers:

    - LCG 0 for signaling radio bearers

    - LCG 1 for **guaranteed bit-rate** (**GBR**) data radio bearers

    - LCG 2 for non-GBR data radio bearers

- Handovers between LTE eNBs

- Power control

- Interference management

The following features or operations are not supported by ns-3.36 LTE eNBs:

- Radio link failure handling

- Interworking with 2G/3G/Wi-Fi technologies

- Connecting with multiple S-GWs or MMEs

- No support in terms of charging UE traffic flows

That's good. We've got to know about ns-3 LTE eNB supporting operations and functionalities for starting our simulations.

Next, we will discuss ns-3 LTE UE supporting operations and functionalities for conducting simulations.

### ns-3 UE operations and functionalities

The ns-3 LTE implementation carefully considers UE implementation to match with ns-3 eNB operations. LTE UE is simulated using ns-3 nodes with LTE UE data plane and control plane protocols. Here, UE data plane protocols handle tasks related to IP packet exchange between UE and the P-GW via a tunnel through eNB. ns-3 LTE UE uses its protocol stack (application, TCP/IP, PDCP, RLC, MAC, and L1) to exchange data traffic between UE and eNB (refer to *Figure 7.3*).

ns-3 LTE UE nodes also simulate control plane protocols (refer to *Figure 7.3*) to handle all control messages exchanged between UE and eNB, and UE and MME. In order to exchange all radio resource-related control signals, relevant messages are exchanged between UE and eNB through the RRC, PDCP, RLC, MAC, and L1 protocol stack. In order to simulate **non-access stratum** (**NAS**) procedures such as **attach**, UE needs direct communication with the LTE EPC node MME. The ns-3 LTE UE attach procedure messages are simulated through the NAS, RRC, PDCP, RLC, MAC, and L1 protocol stack:

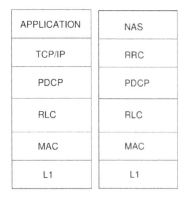

Figure 7.3 – ns-3 UE data plane and control plane protocols

Mainly, ns-3 UE handles the following major operations matching with LTE standards:

- Random access procedures.
- LTE attach procedure.
- Sending UE signal quality measurement reports to eNBs for handover management. ns-3 UE mainly sends the following two measurement reports: **Reference Signal Received Power** (**RSRP**), which is receiving power from a specific eNB, and the **Reference Signal Received Quality** (**RSRQ**), which includes channel interference and thermal noise.
- In the uplink, two types of **Channel Quality Indicators** (**CQIs**) are computed: one is based on **Sounding Reference Signals** (**SRS**) that are periodically sent by the UE to eNBs. The other is based on the actual transmitted data using PUSCH.
- Cell searching for sending connection requests to eNB. During cell search using the measured RSRP, the UE PHY identifies a list of cells, including the cell ID and its average RSRP.
- Cell evaluation: From the list of cells, UE can select the best cell (based on maximum RSRP) to join the cellular network.
- Radio link failure is modeled.

The ns-3.36 LTE UE implementation has the following limitations:

- **EPS Connection Management** (**ECM**) and **EPS Mobility Management** (**EMM**) are not modeled completely
- NAS does not support multiple operators and cell selection groups
- Does not support location update/paging procedures in UE idle mode

That's great! We have discussed ns-3 LTE E-UTRAN supporting protocols, nodes, and their operations. In the next section, we will discuss ns-3 supporting LTE EPC network details for planning our simulations.

### ns-3.36 LTE EPC supporting nodes and their operations

The ns-3 LTE EPC model includes S-GW, P-GW, and MME nodes and their protocols and interfaces. It mainly helps to simulate an end-to-end traffic flow between UE and P-GW. In LTE EPC networks, one of the major objectives is to ensure QoS requirements for UE traffic flows. In LTE architectures, it is ensured by establishing EPS bearers for handling UEs traffic flows. An EPS bearer corresponds to UE and the P-GW end-to-end QoS profile, such as minimum and maximum guaranteed throughputs, maximum tolerable latencies, and maximum allowed packet loss percentages. ns-3 LTE EPC supports only the following two types of bearers for QoS traffic simulations:

- **GBR bearers**: ns-3 LTE RAN and EPC considers GBR parameters while establishing EPS bearers. Hence, ns-3 eNBs can allocate a suitable number of radio resources for ensuring QoS using a priority scheduler. Mainly, users can configure the maximum bit rate and minimum bit rate parameters of GBR bearers. It helps to simulate QoS applications such as VoIP and VoLTE. However, in current ns-3.36 LTE simulations, there is no provision for managing resource allocation at S1-U and S5 interfaces.

- **Non-GBR bearers**: In the case of non-GBR, there is no QoS support from the LTE network. Hence, LTE RAN and EPC components need not reserve any resources at the time of bearer establishment. Non-GBR bearers help to simulate web browsing and file transfer applications.

That means the LTE network should ensure GBR QoS traffic requirements of the UE based on the bearers between the following links, such as the UE and eNB radio link, the eNB and S-GW transmission link, and the S-GW and P-GW transmission link. In the ns-3 LTE EPC implementation, UE flows' QoS requirements are ensured between UE and eNB by configuring the priority scheduler. However, between eNBs and P-GW, it is assumed that abundant link capacity is available to meet the QoS requirements of flows.

We can understand an ns-3 LTE EPS bearer as a logical channel setup that comprises the following sub-bearers to ensure the QoS requirements of traffic flow (source IP, source port, destination IP, and destination port):

- A unique radio bearer is set up between UE and eNB with a unique **Radio Bearer Identifier (RBID)** for each EPS bearer

- A unique S1 bearer is set up between eNB and the S-GW with a unique S1 bearer ID for each EPS bearer

- A unique S5 bearer is set up between the S-GW and the P-GW with a unique S5 bearer ID for each EPS bearer

LTE nodes (eNBs, S-GW, and P-GW) match these bearer IDs and map them to corresponding EPS bearers to ensure the QoS requirements of traffic flows are met.

After understanding the major objectives of ns-3 LTE EPC, let's discuss each of the ns-3 supporting LTE EPC node's functionalities:

- **ns-3 P-GW**: ns-3 P-GW mainly handles IP addressing to UE, forwarding traffic flows between S-GW and the internet, routing, and QoS enforcement only. In LTE simulations, P-GW is considered a gateway router to reach the internet. Hence, in our simulation, internet application servers should set up their default gateway as a P-GW IP address.

  In order to ensure the QoS of traffic flows, ns-3 P-GW matches each of the downlink traffic flows and maps them to an EPS bearer. Hence, using the EPS bearer ID identifies the S5 bearer ID. Then, the downlink flow can be forwarded to the respective S-GW.

  However, ns-3 P-GW does not handle the following tasks:

  - No support in terms of charging UEs traffic flows

  - UE policies enforcement

  - Interworking with other mobile networks such as 2G/3G and Wi-Fi networks

- **ns-3 S-GW**: ns-3 S-GW mainly forwards UE traffic flows to ns-3 P-GW. It also helps in ensuring the QoS of the traffic flow by mapping S1 bearer IDs to radio bearer IDs, and vice versa. Hence, it checks each downlink flow and maps the S1 bearer ID to the RBID, and similarly, the uplink flow RBID maps to the S1 bearer ID.

  However, ns-3 S-GW has the following limitations:

  - No mobility support between S-GWs

  - No load balancing support between S-GWs

  - No support in terms of charging UE traffic flows

  Till this point, we have only discussed ns-3 LTE EPC major data plane nodes details. Next, we will discuss ns-3 LTE EPC major control plane node details.

- **ns-3 MME**: ns-3 MME mainly handles NAS control signaling for simulating the attach procedure between UE and MME. It also helps during handovers between eNBs to configure UE traffic path switching and bearer management.

  However, ns-3 MME does not handle the following:

  - UE authentication

  - Traffic security

  - Paging and tracking area update

  - MME load balancing

Overall, ns-3 LTE EPC simulations do not support the following use cases:

- Roaming between two different mobile networks
- Interworking/handovers with different radio access technologies such as 2G, 3G, and Wi-Fi
- Setting up MME pool areas
- Setting up S-GW pool areas
- Load balancing between S-GWs (data plane)
- Load balancing between MMEs (control plane)
- Policy control, enforcement, and charging
- Tracking area updates and paging
- The focus of the ns-3 EPC model implementation is limited to simulations of active users in ECM-connected mode only

Well done. We have discussed ns-3 supporting LTE E-UTRAN and EPC components and their protocols, operations, and functionalities. In the next section, we will discuss ns-3 supporting interfaces to interconnect ns-3 LTE E-UTRAN and EPC.

## ns-3 LTE architecture supporting major interfaces

ns-3 LTE simulations mainly support the following interfaces to interconnect ns-3 LTE E-UTRAN components and EPC components:

- S1-AP
- S1-U and S5
- X2

Let's look at these in a bit more detail:

- **ns-3 S1-AP**: The S1-AP interface provides control plane interaction between the ns-3 eNB and the ns-3 MME. It handles the following message exchanges in ns-3 LTE simulations for handling the UE attach procedure (INITIAL UE and CONTEXT) and handover procedures (PATH SWITCH):

  - INITIAL UE MESSAGE
  - INITIAL CONTEXT SETUP REQUEST
  - INITIAL CONTEXT SETUP RESPONSE
  - PATH SWITCH REQUEST
  - PATH SWITCH REQUEST ACKNOWLEDGE

- **ns-3 S1-U and S5**: These interfaces are useful for handling initial UE attachments, specifically for handling the establishment of the S1-U and S5 bearers. The S5 interface provides control plane interaction between S-GW and P-GW for establishing UEs EPS bearers. The following message exchange is simulated:

    - CREATE SESSION REQUEST

    - CREATE SESSION RESPONSE

    When source eNB receives a PATH SWITCH REQUEST S1-AP message from MME, the following messages are simulated in ns-3 LTE simulations to switch the S1-U bearers from the source eNB to the target eNB:

    - MODIFY BEARER REQUEST

    - MODIFY BEARER RESPONSE

    - DELETE SESSION REQUEST

    - DELETE SESSION RESPONSE

    - DELETE BEARER COMMAND

    - DELETE BEARER REQUEST

    - DELETE BEARER RESPONSE

- **X2 interface**: X2 interfaces are useful to exchange load, interference, and handover-related information between eNBs. The standard LTE X2 interface is useful for exchanging the following details:

    - Load management-related messages for implementing load balancing between eNBs

    - Interference-related messages are exchanged between eNBs for interference coordination between neighbor eNBs (or cells)

    - Handling UE mobility between eNBs using handover message exchanges such as handover request, handover response, buffer status, and sequence numbers transfer

    The X2 interfaces need to be set up between eNBs to simulate handover and interference management scenarios in ns-3 LTE networks. However, the ns-3.36 implementation does not support load-balancing algorithms.

That's perfect. Now, we are ready to learn how to simulate LTE network use cases using ns-3. In the next section, let's discuss how to set up and configure various ns-3-supporting LTE E-UTRAN and EPC components to implement our simulation scenarios.

# Learning how to set up and configure various LTE nodes in ns-3

In this section, we will discuss how to set up an LTE network using ns-3-supporting RAN and EPC nodes, interfaces, and protocols. Mainly, we will discuss how to set up various LTE nodes and configure their RAN and EPC features in ns-3.

## ns-3 LTE RAN nodes setup and their configuration

Let's start with creating LTE eNBs and 100 LTE UEs using ns-3 nodes with the following code snippet:

```
#include "ns3/lte-module.h"
  NodeContainer eNodeBs;
  NodeContainer UEs;
  eNodeBs.Create (3);
  UEs.Create (100);
```

Let's configure LTE eNB and UEs operating UL and DL frequency bands using the following code:

```
lteHelper->SetEnbDeviceAttribute ("DlEarfcn", UintegerValue
(100));
lteHelper->SetEnbDeviceAttribute ("UlEarfcn", UintegerValue
(18100));
```

Let's see how to configure LTE operating UL and DL bandwidth in terms of RBs using the following line of code:

```
uint8_t RBs = 100;
lteHelper->SetEnbDeviceAttribute ("DlBandwidth", UintegerValue
(RBs));
lteHelper->SetEnbDeviceAttribute ("UlBandwidth", UintegerValue
(RBs);
```

Here, we configure both uplink and downlink transmission bandwidths the same. It is also possible to configure different bandwidths for uplink and downlink transmissions.

Let's configure eNB and UEs transmission power and uplink power control using the following lines of code:

```
Config::SetDefault ("ns3::LteEnbPhy::TxPower", DoubleValue
(43.0));
Config::SetDefault ("ns3::LteUePhy::TxPower", DoubleValue
```

```
(20.0));
Config::SetDefault ("ns3::LteUePhy::EnableUplinkPowerControl",
BooleanValue (true));
```

> **Important note**
> In case we need to configure LTE small cells, such as Femto or Pico, the appropriate power can be configured.

Similar to Wi-Fi network simulations, LTE RAN networks can be configured using realistic propagation loss models and error models using the following lines of code:

```
lteHelper->SetAttribute ("PathlossModel", StringValue (
,"ns3::BuildingsPropagationLossModel"));
Config::SetDefault
("ns3::LteSpectrumPhy::CtrlErrorModelEnabled", BooleanValue
,(false));
Config::SetDefault
("ns3::LteSpectrumPhy::DataErrorModelEnabled", BooleanValue
,(false));
Config::SetDefault ("ns3::LteEnbRrc::SrsPeriodicity",
UintegerValue (srsPeriodicity));
```

### Setting up three sector eNBs using ns-3 antenna module features

In ns-3 LTE simulation, it is possible to configure eNBs with or without sectors. By default, eNBs do not have any sectors. For example, we want to create three sectors of LTE eNBs with the following specifications using the ns-3 supporting antenna models:

- Configure three sectors and set the sector with a unique operating cell ID
- Configure the cosine antenna model to set the eNB antenna operating model
- Configure 3 sector orientations at 0, 120, and 240 degrees, respectively
- Configure equal beamwidth for each of the sectors
- Configure maximum gain of antennas

Let's see how to configure the first sector using the following ns-3 lines of code:

```
lteHelper->SetEnbAntennaModelType
("ns3::CosineAntennaModel");
lteHelper->SetEnbAntennaModelAttribute ("Orientation",
DoubleValue (0));
```

```
   lteHelper->SetEnbAntennaModelAttribute
("HorizontalBeamwidth", DoubleValue (100));
   lteHelper->SetEnbAntennaModelAttribute ("MaxGain",
DoubleValue (0.0));
   enbDevs.Add ( lteHelper->InstallEnbDevice (threeSectorNodes.
Get (0)));
```

Similarly, let's configure the second sector of the eNB using the following code snippet:

```
   lteHelper->SetEnbAntennaModelType
("ns3::CosineAntennaModel");
   lteHelper->SetEnbAntennaModelAttribute ("Orientation",
DoubleValue (360/3));
   lteHelper->SetEnbAntennaModelAttribute
("HorizontalBeamwidth", DoubleValue (100));
   lteHelper->SetEnbAntennaModelAttribute ("MaxGain",
DoubleValue (0.0));
   enbDevs.Add ( lteHelper->InstallEnbDevice (threeSectorNodes.
Get (1)));
```

Finally, let's configure the third sector of the eNB using the following code snippet:

```
   lteHelper->SetEnbAntennaModelType
("ns3::CosineAntennaModel");
   lteHelper->SetEnbAntennaModelAttribute ("Orientation",
DoubleValue (2*360/3));
   lteHelper->SetEnbAntennaModelAttribute
("HorizontalBeamwidth", DoubleValue (100));
   lteHelper->SetEnbAntennaModelAttribute ("MaxGain",
DoubleValue (0.0));
   enbDevs.Add ( lteHelper->InstallEnbDevice (threeSectorNodes.
Get (2)));
```

> **Important note**
> In the case of setting up three sector antenna eNBs, it is necessary to set up three ns-3 eNB nodes, and all three nodes should be placed at the same location.

Next, let's see how to configure LTE MIMO in ns-3 simulations.

### Configuring the MIMO feature in ns-3 LTE simulations

ns-3 LTE supports the following seven transmission modes:

- **Transmission Mode 1**: SISO

- **Transmission Mode 2**: MIMO Tx Diversity

- **Transmission Mode 3**: MIMO Spatial Multiplexity Open Loop

- **Transmission Mode 4**: MIMO Spatial Multiplexity Closed Loop

- **Transmission Mode 5**: MIMO Multi-User

- **Transmission Mode 6**: Closer loop single-layer precoding

- **Transmission Mode 7**: Single antenna port 5

Let's see how to configure these modes in ns-3. For example, let's configure the SISO mode using the following code snippet:

```
Config::SetDefault ("ns3::LteEnbRrc::DefaultTransmissionMode",
UintegerValue (0));
```

Let's see how to configure MIMO transmission diversity 1 using the following code snippet:

```
Config::SetDefault ("ns3::LteEnbRrc::DefaultTransmissionMode",
UintegerValue (1));
```

Let's see how to configure MIMO transmission diversity 2 using the following code snippet:

```
Config::SetDefault ("ns3::LteEnbRrc::DefaultTransmissionMode",
UintegerValue (2));
```

Well done. We have discussed all basic LTE E-UTRAN node configurations in ns-3. Let's see how to configure various LTE network management algorithms in the next section.

## Configuring ns-3 resource scheduling, handover, and interference control algorithms

Let's set the **round-robin scheduler** for LTE networks to simulate **radio resource scheduling** using the following code:

```
lteHelper->SetSchedulerType ("ns3::RrFfMacScheduler");
```

We will see details of various ns-3 LTE supporting schedulers in upcoming chapters.

Now, let's see how to configure an **ns-3 supporting handover algorithm** for LTE network mobility scenario simulations:

```
lteHelper->SetHandoverAlgorithmType
("ns3::A2A4RsrqHandoverAlgorithm");
lteHelper->SetHandoverAlgorithmAttribute
("ServingCellThreshold",
UintegerValue (30));
lteHelper->SetHandoverAlgorithmAttribute
("NeighbourCellOffset",
UintegerValue (1));
```

Here, we configured A2A4Rsrq handover algorithms and their parameters. We will see details of various ns-3 LTE supporting handover algorithms in upcoming chapters.

Next, let's see how to configure **interference management algorithms** in ns-3:

```
lteHelper->SetFfrAlgorithmType ("ns3::LteFrHardAlgorithm");
```

We will see details of various ns-3 LTE-supporting interference management algorithms in upcoming chapters.

### Installing ns-3 LTE protocol stacks on UEs and eNBs, and attaching UEs in the LTE network

Let's start with installing the LTE protocol stack on these nodes and collect the respective network devices:

```
Ptr<LteHelper> lteHelper = CreateObject<LteHelper> ();
NetDeviceContainer enbDevs;
NetDeviceContainer ueDevs;
enbDevs = lteHelper->InstallEnbDevice (eNBs);
ueDevs = lteHelper->InstallUeDevice (UEs);
```

UE attach means connecting UEs to the LTE network. In ns-3, we can connect LTE UEs in the following two ways to eNBs:

- Signal strength-based cell selection by UEs to attach.

  Let's see how to configure UEs to connect to the strongest signal-receiving cell using the following line of code:

  ```
  lteHelper->Attach(ueDevs); // attach one or more UEs to
  the strongest cell
  ```

- Manual cell selection by UEs to attach.

    Let's see how to attach UEs to specific eNBs using the following code:

    ```
    lteHelper->Attach(ueDevs, enbDevs);
    ```

### ns-3 LTE supports home eNBs attach procedure

An interesting use case of the initial cell selection process is to set up a simulation environment with a **closed subscriber group** (**CSG**) deployed by operators for home eNBs. Here, we need to restrict which UEs to connect to home eNBs such as Femtos. In the ns-3 LTE module, it can be simulated by configuring the same CSG IDs with home eNBs and their UEs.

Let's see how to do it in ns-3 LTE simulations using the following lines of code:

1.  First, enable CSG and configure home eNBs' CSG ID to 1 using the following lines of code:

    ```
    lteHelper->SetEnbDeviceAttribute ("CsgId", UintegerValue
    (1));
    lteHelper->SetEnbDeviceAttribute ("CsgIndication",
    BooleanValue (true));
    ```

2.  Next, configure the UE CSG ID to 1 using the following line of code:

    ```
    lteHelper->SetUeDeviceAttribute ("CsgId", UintegerValue
    (1));
    ```

3.  After CSG configuration only, we need to install the LTE protocol stack on eNBs and UEs:

    ```
    NetDeviceContainer csgEnbs = lteHelper->InstallEnbDevice
    (eNbNodes);
    NetDeviceContainer csgUes = lteHelper->InstallUeDevice
    (UeNodes);
    lteHelper->Attach (csgUes);
    ```

> **Important note**
>
> CSG attach simulations work with only automatic cell selection. It is also possible to disable CSG restriction at eNB by configuring `CsgIndication` as `false` (the default value for eNB). Hence, any UE can connect to the eNB. Include the following lines of code to enable `CsgIndication`: (`lteHelper->SetEnbDeviceAttribute ("CsgIndication", BooleanValue (true));`).

That's great we have completed discussing how to set up and configure LTE E-UTRAN components using ns-3 LTE code snippets. In the next section, we will discuss how to connect LTE E-UTRAN with EPC for simulating complete LTE EPS scenarios.

## ns-3 LTE EPC node setup and configuration

Till now, we discussed only LTE RAN-related configurations. In order to simulate UE IP traffic flow between RAN and EPC and set up EPS bearers for QoS traffic simulations, it is necessary to set up and configure ns-3 LTE EPC nodes.

Let's start with how to set up ns-3 LTE EPC using the following lines of code:

```
Ptr<LteHelper> lteHelper = CreateObject<LteHelper> ();
Ptr<PointToPointEpcHelper> epcHelper =
CreateObject<PointToPointEpcHelper> ();
lteHelper->SetEpcHelper (epcHelper);
```

With these lines of code, all ns-3 EPC nodes (such as the S-GW, P-GW, and MME) are created for LTE simulations. However, internet services are still not available. We need to set up internet nodes that should get connected to the ns-3 P-GW node.

For simulations, we need to connect internet service simulation ns-3 nodes, and ns-3 P-GW nodes should be connected using P2P channels as follows.

Let's start with configuring the ns-3 P-GW node object using the following line of code:

```
Ptr<Node> pgw = epcHelper->GetPgwNode ();
```

Now, create an internet node and connect it to the P-GW via the P2P channel using the following lines of code:

```
NodeContainer InternetNode;
InternetNode.Create (1);
InternetStackHelper internet;
internet.Install (InternetNode);
PointToPointHelper p2p;
p2ph.SetDeviceAttribute ("DataRate", DataRateValue (DataRate
("100Gb/s")));
p2ph.SetDeviceAttribute ("Mtu", UintegerValue (1500));
p2ph.SetChannelAttribute ("Delay", TimeValue (Seconds
(0.010)));
NetDeviceContainer internetDevices = p2ph.Install (pgw,
InternetNode);
```

In order to exchange traffic between the internet node and the LTE EPC network, we need to configure a static route to the LTE EPC network on the internet node using the following lines of code:

```
Ipv4AddressHelper ip;
ip.SetBase ("1.0.0.0", "255.0.0.0");
Ipv4InterfaceContainer interfaces = ipv4h.Assign
(internetDevices);
Ipv4Address internetHostAddr = interfaces.GetAddress (1);
Ipv4StaticRoutingHelper routing;
Ptr<Ipv4StaticRouting> internetHostrouting;
internetHostrouting = routing.GetStaticRouting (InternetNode-
>GetObject<Ipv4> ());
internetHostrouting->AddNetworkRouteTo (epcHelper-
>GetEpcIpv4NetworkAddress (),
Ipv4Mask ("255.255.0.0"), 1);
```

On the other hand, UEs also needed to be configured with static IP routing to reach the P-GW and internet. Let's configure static routing on UEs using the following lines of code:

```
InternetStackHelper internet;
internet.Install (ueNodes);
for (uint32_t u = 0; u < ueNodes.GetN (); ++u)
{
   Ptr<Node> ue = ueNodes.Get (u);
   Ptr<NetDevice> ueLteDevice = ueLteDevs.Get (u);
   Ipv4InterfaceContainer ueIpIface;
   ueIpIface = epcHelper-
>AssignUeIpv4Address  (NetDeviceContainer (ueLteDevice));
   Ptr<Ipv4StaticRouting> ueStaticRouting;
   ueStaticRouting = ipv4RoutingHelper.GetStaticRouting (ue-
>GetObject<Ipv4> ());
   ueStaticRouting->SetDefaultRoute (epcHelper-
>GetUeDefaultGatewayAddress (), 1);
 }
```

Till this point, we discussed how to connect ns-3 LTE E-UTRAN and EPC. In the next section, we will discuss how to simulate EPS bearers for QoS simulation setup scenarios.

### Configuration of EPS bearers on UE to simulate QoS traffic flows

Let's see, for example, how to configure EPS bearers for a UE with specific GBR configurations (512 Kbps for UL and DL flows separately) in ns-3 simulations using the following code snippet:

```
Ptr<NetDevice> ueDevice = UEdevs.Get (u);
GbrQosInformation qos;
qos.gbrDl = (512+32)*100*8;
qos.gbrUl = (512+32)*100*8;
qos.mbrDl = qos.gbrDl;
qos.mbrUl = qos.gbrUl;
enum EpsBearer::Qci q = EpsBearer::GBR_CONV_VOICE;
EpsBearer bearer (q, qos);
lteHelper->ActivateDedicatedEpsBearer (ueDevice, bearer,
EpcTft::Default ());
```

That's great. We have learned how to set up an LTE network and install the necessary protocols for executing our simulations using ns-3. Before starting your first LTE simulation, let's summarize common ns-3 LTE simulation configurations as follows:

- ns-3 LTE simulations can be configured only in FDD operating mode. That means for uplink and downlink transmissions, separate bandwidths need to be configured.

- In order to comply with LTE standard implementation, the ns-3 LTE module also uses OFDMA for downlink transmissions (eNB to UE), and SC-FDMA for uplink transmissions (UE to eNB).

- The ns-3 LTE operating frame duration is 10 ms and the subframe duration is 1 ms.

- ns-3 LTE simulations consider the spectrum in terms of RBs only. Similar to the LTE standard, ns-3 LTE simulations uplink or downlink bandwidths can be configured with 6, 15, 50, 75, and 100 RBs. By default, ns-3 round-robin scheduling algorithms are configured for simulations.

- There is no default handover or interference management enabled for ns-3 LTE simulations.

- ns-3 LTE EPC must be configured to simulate complete ns-3 LTE EPS scenarios.

- Internet application servers should be connected to ns-3 P-GW to simulate internet scenarios. By default, the internet is not available in ns-3 LTE simulations.

- To simulate the QoS scenario, EPS bearers should be configured in ns-3 LTE simulations.

Finally, the wait is over. Let's start our first LTE simulation in the next section.

# Simulating LTE RAN and EPC networks using ns-3 key building blocks

Having discussed ns-3 supporting LTE features and how to set up and configure various LTE network nodes, we can start setting up LTE/4G network simulations in a hands-on activity in this section. In this hands-on activity, we focus on setting up complete LTE EPS (RAN and EPC) and handling the following tasks:

- As part of LTE RAN, deploy two eNBs with a distance of 500 meters between them
- Deploy a minimum of 20 LTE UEs near each eNB
- Configure eNBs and UE transmission power as 46 dB and 20 dB, respectively
- Configure UL and DL bandwidths to 25 RBs
- Configure propagation loss model and error models
- Configure LTE RRC parameter SRS based on the number of UEs to connect with eNB
- Install LTE UE and eNB protocol stacks
- Set up EPC and connect with RAN
- Set up an internet host and connect with a P-GW using a 10 Gbps P2P link
- Test automatic UE attach and manual attach procedures to connect with the LTE network (track which UEs are connected to which eNBs)
- Test by default how many UEs can be connected to an eNB
- Test UL and DL throughputs of UEs separately using UDP applications
- Test how to increase cell capacity in case more UEs connect to the network

## Setting up and configuring the LTE EPS network hands-on activity using ns-3

Let's start our LTE EPS simulation in `pkt-lte-epc.cc` by importing all necessary ns-3 LTE modules for setting up our LTE EPS network simulation:

```
#include "ns3/lte-helper.h"
#include "ns3/epc-helper.h"
#include "ns3/core-module.h"
#include "ns3/network-module.h"
#include "ns3/ipv4-global-routing-helper.h"
#include "ns3/internet-module.h"
```

```
#include "ns3/mobility-module.h"
#include "ns3/lte-module.h"
#include "ns3/applications-module.h"
#include "ns3/point-to-point-helper.h"
#include "ns3/config-store.h"
#include "ns3/flow-monitor-module.h"
using namespace ns3;
```

Handle the tracking of UE connection using the following callback function. When a UE is successfully connected to an eNB, it prints the UE **International Mobile Subscriber Identity** (**IMSI**), **Radio Network Temporary Identifier** (**RNTI**), and connected **Cell Identifier** (**Cell ID**) details:

```
void UEsConnectionStatus (std::string context,
                          uint64_t imsi,
                          uint16_t cellid,
                          uint16_t rnti)
{
  std::cout << " UE IMSI " << imsi
            << " connected to CellId " << cellid
            << " with RNTI " << rnti<<"
Time:"<<Simulator::Now()
            << std::endl;
}
```

Start the main() function by creating the necessary ns-3 nodes for setting up LTE eNBs and UEs:

```
int
main (int argc, char *argv[])
{
  uint16_t nUEs = 40;
  double simTime = 2.0;
  NodeContainer UEs;
  NodeContainer eNBs;
  eNBs.Create(2);
  UEs.Create(nUEs);
```

After creating the necessary LTE eNB and UE nodes, deploy them as per the hands-on activity; in this activity, eNBs are separated by a 500 m distance. Deploy half of the UEs close to one eNB and the rest of the UEs near another eNB. Moreover, in this simulation, eNBs and UEs are placed at constant positions only. We will discuss mobility use case simulations in the upcoming chapters. We use `ListPositionAllocator` to set up this topology using the following ns-3 lines of code:

```
  Ptr<ListPositionAllocator> positionAlloc =
CreateObject<ListPositionAllocator> ();
  positionAlloc->Add (Vector(0, 0, 0));
  positionAlloc->Add (Vector(500, 0, 0));
  for (uint16_t i = 0; i < nUEs/2; i++)
    {
      positionAlloc->Add (Vector(i+10, 0, 0));
    }
  for (uint16_t i = nUEs/2; i < nUEs; i++)
    {
      positionAlloc->Add (Vector(500+i, 0, 0));
    }
  MobilityHelper mobility;
  mobility.
SetMobilityModel("ns3::ConstantPositionMobilityModel");
  mobility.SetPositionAllocator(positionAlloc);
  mobility.Install(eNBs);
  mobility.Install(UEs);
```

After setting up the LTE RAN topology as per the hands-on activity using the following ns-3 lines of code, let's configure LTE RAN node parameters such as eNB and UEs transmission power, enabling LTE data, and control-channel error modeling:

```
  Config::SetDefault ("ns3::LteEnbPhy::TxPower", DoubleValue
(46.0));
  Config::SetDefault ("ns3::LteUePhy::TxPower", DoubleValue
(20.0));
  Config::SetDefault
("ns3::LteSpectrumPhy::CtrlErrorModelEnabled",
BooleanValue(true));
  Config::SetDefault
("ns3::LteSpectrumPhy::DataErrorModelEnabled",
BooleanValue(true));
```

Configure LTE SRS periodicity to test the number of UEs that are supported in ns-3 LTE simulations. We initialized it with 20 to support 20 UEs, and we change this value during our simulation setup testing to increase the number of UEs:

```
Config::SetDefault ("ns3::LteEnbRrc::SrsPeriodicity",
UintegerValue (20));
```

Till this point, we configured general LTE RAN-specific configurations. Next, we see how to set up LTE EPC and connect with LTE RAN using the following lines of code. As a result of this configuration, LTE EPC nodes are created, and LTE RAN is connected with the EPC using ns-3-supporting interfaces such as S1-U, S5, S1-AP, and S11:

```
Ptr<LteHelper> lteHelper = CreateObject<LteHelper> ();
Ptr<PointToPointEpcHelper> epcHelper =
CreateObject<PointToPointEpcHelper> ();
lteHelper->SetEpcHelper (epcHelper);
```

Next, we see how to configure a path loss model for our LTE simulation:

```
lteHelper->SetAttribute ("PathlossModel", StringValue
("ns3::FriisPropagationLossModel"));
```

Then, configure radio and spectrum resources in our simulation setup in terms of RBs. Initially, we set UL and DL RBs to 25, but during testing, we change these values:

```
lteHelper->SetEnbDeviceAttribute ("DlBandwidth",
UintegerValue (25));
lteHelper->SetEnbDeviceAttribute ("UlBandwidth",
UintegerValue (25));
```

After setting up and configuring LTE RAN with EPC, we can install LTE protocol stacks on eNBs and UEs using the following lines of code:

```
NetDeviceContainer eNBdevs = lteHelper->InstallEnbDevice
(eNBs);
NetDeviceContainer UEdevs = lteHelper->InstallUeDevice (UEs);
```

Although setting up EPC is done, to simulate internet service access from the LTE network, we must create an internet host and connect with the P-GW using the P2P channel explicitly, as follows.

First, set up an internet host to install our testing application using the following lines of code:

```
NodeContainer internetHost;
internetHost.Create (1);
```

```
InternetStackHelper internet;
internet.Install (internetHost);
```

Now, connect the internet host with the P-GW using a P2P channel of 10 Gbps speed using the following lines of code:

```
Ptr<Node> ihost = internetHost.Get (0);
Ptr<Node> pgw = epcHelper->GetPgwNode ();
 PointToPointHelper p2p;
 p2p.SetDeviceAttribute ("DataRate", DataRateValue (DataRate
("10Gb/s")));
 p2p.SetDeviceAttribute ("Mtu", UintegerValue (1500));
 p2p.SetChannelAttribute ("Delay", TimeValue (Seconds
(0.010)));
 NetDeviceContainer inetDevs = p2p.Install (pgw,ihost);
```

In ns-3 LTE simulations, we must configure the 1.0.0.0/8 subnet for the P-GW and internet host interfaces. Then, it is also necessary to set up a static route on the internet host to reach ns-3 LTE UE, as follows:

```
Ipv4AddressHelper ip;
ip.SetBase ("1.0.0.0", "255.0.0.0");
Ipv4InterfaceContainer ipinterfs = ip.Assign (inetDevs);
Ipv4Address hostaddr = ipinterfs.GetAddress (1);
Ipv4StaticRoutingHelper iproute;
Ptr<Ipv4StaticRouting> ihostroute  = iproute.GetStaticRouting
(ihost->GetObject<Ipv4> ());
 ihostroute->AddNetworkRouteTo (Ipv4Address ("7.0.0.0"),
Ipv4Mask ("255.0.0.0"), 1);
```

Besides this, it is also necessary to configure all UE IP addresses using the ns-3 EpcHelper application and set up a default gateway address for all UEs explicitly using static routing:

```
internet.Install (UEs);
Ipv4InterfaceContainer ueIPs;
ueIPs = epcHelper->AssignUeIpv4Address (NetDeviceContainer
(UEdevs));
for (uint32_t i = 0; i < UEs.GetN (); ++i)
  {
    Ptr<Node> ue = UEs.Get (i);
```

```
        Ptr<Ipv4StaticRouting> ueroute = iproute.GetStaticRouting
(ue->GetObject<Ipv4> ());
        ueroute->SetDefaultRoute (epcHelper-
>GetUeDefaultGatewayAddress (), 1);
    }
```

Till this point, we have done all necessary ns-3 LTE RAN and EPC configurations to simulate internet service access.

Now, let's attach LTE UEs to the LTE network manually – specifically, connect half of the UEs to eNB-1 and the remaining UEs to the eNB-2 using the following lines of code:

```
for (uint16_t i = 0; i < nUEs/2; i++)
  {
    lteHelper->Attach (UEdevs.Get(i), eNBdevs.Get(0));
  }
for (uint16_t i = nUEs/2; i < nUEs; i++)
  {
    lteHelper->Attach (UEdevs.Get(i), eNBdevs.Get(1));
  }
```

It is also possible to simulate the automatic UE attach procedure to connect with the strongest cell in terms of UEs receiving signal strength using the following lines of code (initially, we comment, but during evaluation, we can uncomment and test it):

```
//  lteHelper->Attach (ueLteDevs);
```

That's all for setting up and configuring complete LTE EPS network simulation. Next, start setting up DL flows (internet host -> UEs) and UL flow (UEs -> internet host) for evaluation of our ns-3 LTE simulation setup.

Let's set up a unique DL flow between each UE and internet host using ns-3 `UdpClient` and ns-3 `PacketSink` applications. In the following loop, we install UE-specific `PacketSink` applications on UEs and respective `UdpClient` applications on the internet host. Here, we configure around 400 Kbps UDP application data rate for each DL flow (internet host to UE) using the following lines of code:

```
uint16_t dlPort = 5001;
ApplicationContainer clientApps;
ApplicationContainer serverApps;
for (uint32_t u = 0; u < UEs.GetN (); ++u)
  {
```

```
        PacketSinkHelper dlPacketSinkHelper
  ("ns3::UdpSocketFactory", InetSocketAddress
  (Ipv4Address::GetAny (), dlPort));
        serverApps.Add (dlPacketSinkHelper.Install (UEs.Get(u)));
        UdpClientHelper dlClient (ueIPs.GetAddress (u), dlPort);
        dlClient.SetAttribute ("Interval", TimeValue
  (Seconds(0.01)));
        dlClient.SetAttribute ("PacketSize", UintegerValue(512));
        dlClient.SetAttribute ("MaxPackets",
  UintegerValue(10000));
        clientApps.Add (dlClient.Install (ihost));
        dlPort++;
      }
    serverApps.Start (Seconds (1.01));
    clientApps.Start (Seconds (1.01));
```

Similarly, set up a unique UL flow between each UE and internet host using the ns-3 UdpClient and ns-3 PacketSink applications. In the following loop, we install UE-specific UdpClient applications on the internet host and the respective PacketSink applications on the internet host. Here, we configure around 400 Kbps UDP application data rate for each UL flow (UE to internet host) using the following lines of code:

```
  ApplicationContainer uclientApps;
  ApplicationContainer userverApps;
  uint16_t ulPort = 16001;
  for (uint32_t u = 0; u < UEs.GetN(); ++u)
    {
      ulPort++;
      PacketSinkHelper ulPacketSinkHelper
  ("ns3::UdpSocketFactory", InetSocketAddress
  (Ipv4Address::GetAny (), ulPort));
      userverApps.Add (ulPacketSinkHelper.Install (ihost));
      UdpClientHelper ulClient (hostaddr, ulPort);
      ulClient.SetAttribute ("Interval", TimeValue
  (Seconds(0.01)));
      ulClient.SetAttribute ("PacketSize", UintegerValue(512));
      ulClient.SetAttribute ("MaxPackets",
  UintegerValue(10000));
      uclientApps.Add (ulClient.Install (UEs.Get(u)));
```

```
   }
userverApps.Start (Seconds (1.01));
uclientApps.Start (Seconds (1.01));
```

We are almost done with our simulation setup and evaluation configurations. Next, to view various LTE simulation results, enable LTE protocol layer-level statistics collection. Then, configure UEs connection setup traces to track UEs connection with eNBs using the following lines of code:

```
lteHelper->EnableTraces ();
Config::Connect ("/NodeList/*/DeviceList/*/LteUeRrc/
ConnectionEstablished",
                MakeCallback (&UEsConnectionStatus));
```

Finally, close main () by setting up a flow monitor to collect flow level stats (such as throughput, delay, and packet loss) and free all simulation resources using the following lines of code:

```
Ptr<FlowMonitor> flowmon;
FlowMonitorHelper flowmonHelper;
flowmon = flowmonHelper.InstallAll ();
Simulator::Stop(Seconds(simTime+10));
Simulator::Run();
flowmon->SerializeToXmlFile ("lte_flowmetrics.xml", true,
true);
Simulator::Destroy();
return 0;
}
```

That's excellent. We have successfully set up our first LTE network for simulations using ns-3. Without waiting, let's now evaluate it thoroughly.

## Evaluation of LTE EPS simulation setup

In this section, we will mainly focus on evaluating the following LTE aspects using the pkt-lte-epc.cc simulation:

- Testing UE manual attachment and automatic attachment procedures
- Exploring the maximum number of UEs supported per eNB in the ns-3.36 LTE module
- Understanding UL and DL flow requirements and configuring suitable bandwidths

## Testing automatic UE attach and manual attach procedures

Let's start with exploring the LTE manual attachment procedure. In this scenario, we mainly focus on which UEs are connected to which eNBs (Cell ID) by tracking UEs' successful attachment (using the *ns-3 Tracing* feature) details such as IMSI, RNTI, and Cell ID.

Start executing our first test scenario using the following commands at the ns-3.36 installation directory:

```
./ns3 run scratch/pkt-lte-epc |tee UEManualAttach
```

Here, we are collecting UE attachment details in the UEManualAttach file.

Let's check how many UEs are connected with each eNB using the following command. In our simulation, each eNB is configured with a unique Cell ID (eNB-1: CellId 1, and eNB-2: CellId 2)

To get the UE attachment, let's execute the following command:

```
cat UEManualAttach |grep 'CellId 1' |wc -l
20
```

That's perfect. As per our ns-3 code, the first 20 UEs should be configured to CellId 1:

```
cat UEManualAttach |grep 'CellId 2' |wc -l
20
```

That's perfect. As per our ns-3 code, another 20 UEs should be configured to CellId 2.

Now, let's check ns-3 LTE UEs auto-attachment procedure using the following command:

```
./ns3 run scratch/pkt-lte-epc |tee UEAutoAttach
```

We collected UE attachment details into the UEAutoAttach file.

Let's check the number of UEs that are connected with each eNB (cell) using the following commands:

```
cat UEAutoAttach |grep 'CellId 1'  |wc -l
20
cat UEAutoAttach |grep 'CellId 2'  |wc -l
20
```

That's perfect. As per the ns-3 LTE UE auto-attachment procedure, UEs must connect with the strongest signal-receiving cell. In our simulation, we placed the first 20 UEs near eNB-1 and another 20 UEs near eNB-2, hence we are observing that each Cell ID has 20 UEs connected.

### Testing default maximum UE attach support with an eNB (or cell)

We have seen that 20 UEs per cell are allowed without any issues. Let's increase the number of UEs to 40 per cell and run the simulation using the following commands:

```
./ns3 run scratch/pkt-lte-epc
msg="too many UEs (21) for current SRS periodicity 20, consider
increasing the value of ns3::LteEnbRrc::SrsPeriodicity",
+0.269000000s 0 file=/home/anil/ns-allinone-3.36/ns-3.36/src/
lte/model/lte-enb-rrc.cc, line=3090
terminate called without an active exception
Command 'build/scratch/ns3.36-pkt-lte-epc-default' died with
<Signals.SIGABRT: 6>.
```

Oh! It crashes, saying there are too many UEs. That means, by default, only 20 UEs are supported per cell. We need to increase SRS periodicity to support more UEs per cell in our simulations. Let's test it in the next scenario.

### Testing how to increase the number of UEs in the simulation

Usually, SRS periodicity-allowed values are 2, 5, 10, 20, 40, 80, 160, and 320 only.

In our simulation, pkt-lte-epc.cc, we initially configured SRS periodicity to 20; we can check that 20 UEs per cell are supported. To support 40 UEs per cell, SRS should be set to 40, and so on. We suggest you try configuring the maximum SRS periodicity to 320 and try to increase the number of UEs per cell. ns-3.36 LTE supports 275 UEs per cell. However, if we increase the number of UEs to more than 275, the simulation again crashes.

### Testing LTE UL and DL bandwidth capacity

Now, we understand how increasing SRS periodicity helps to increase the amount of UEs per cell. However, we did not test their UL and DL data rates with a bandwidth of RBs only.

Let's execute the following commands to test a simulation configuration of 40 UEs per cell (per UE two flows), SRS periodicity to 40, UL RBs to 25, and DL RBs to 25:

```
./ns3 run scratch/pkt-lte-epc
python3 scratch/flowmon-parse-results.py lte_flowmetrics.xml
...
FlowID: 159 (UDP 1.0.0.2/49227 --> 7.0.0.76/5075)
TX bitrate: 432.39 kbit/s
RX bitrate: 237.12 kbit/s
Mean Delay: 333.46 ms
```

```
Packet Loss Ratio: 4.31 %
FlowID: 160 (UDP 1.0.0.2/49228 --> 7.0.0.77/5076)
TX bitrate: 432.39 kbit/s
RX bitrate: 237.12 kbit/s
Mean Delay: 341.82 ms
Packet Loss Ratio: 4.31 %
```

We can observe that a lot of UE is suffering from poor throughput of fewer than 250 Kbps. Let's check whether increasing UL and DL bandwidths to 50 RBs (change UL and DL RBs in `pkt-lte-epc.cc`) will help UEs to get better throughput:

```
./ns3 run scratch/pkt-lte-epc
python3 scratch/flowmon-parse-results.py lte_flowmetrics.xml
...
FlowID: 159 (UDP 1.0.0.2/49227 --> 7.0.0.76/5075)
TX bitrate: 432.39 kbit/s
RX bitrate: 433.98 kbit/s
Mean Delay: 35.03 ms
Packet Loss Ratio: 0.00 %
FlowID: 160 (UDP 1.0.0.2/49228 --> 7.0.0.77/5076)
TX bitrate: 432.39 kbit/s
RX bitrate: 433.82 kbit/s
Mean Delay: 34.54 ms
Packet Loss Ratio: 0.00 %
```

That's a good observation. By increasing UL and DL operating bandwidths, UEs experience improved throughput in its UL and DL transmissions.

> **Important note**
>
> We evaluated our LTE simulation results using a flow monitor for viewing UE-to-internet-host flow level results. We can also inspect LTE protocol-level stats between UE and eNBs by viewing respective PDCP, RLC, MAC, PHY layer UL, and DL flow statistics. These files are generated at the end of the simulation – for example, DL RLC layer stats in `DlRlcStats.txt` and UL RLC layer stats in `UlRlcStats.txt`. We suggest you go through the ns-3 LTE documentation for more details about stats files.

## Summary

In this chapter, you were introduced to LTE architecture, major LTE network components, their protocol stacks, and interfaces for quickly understanding ns-3 LTE simulations. Mainly, we highlighted various ns-3 LTE network components' limitations, and non-supporting protocols and features. Hence, it helps you to avoid mistakes while conducting LTE simulations. Later, we discussed how to set up and configure ns-3 LTE network components such as eNBs, UEs, and EPC components with their protocols and necessary algorithms for conducting any LTE simulations. Finally, you were introduced to a basic hands-on activity using ns-3 to easily set up and configure any LTE end-to-end network simulations. As part of evaluating the LTE network simulation, you learned how to attach UEs to eNBs, configuring the maximum UEs per eNB, and increasing UL/DL bandwidth RBS to meet UEs throughput requirements.

After learning the basics of ns-3 LTE network simulations, in the next chapter, you will learn details of ns-3 modeling of LTE radio resource scheduling and handover implementation for simulating LTE radio resource scheduling and mobility evaluation scenarios.

# 8

# Researching LTE Network Radio Resource Management and Mobility Management Using ns-3

After learning about ns-3 supporting LTE architecture and features in *Chapter 7*, in this chapter, we focus on learning how to simulate LTE mobile network use cases such as spectrum and radio resource allocation to UEs, ensuring QoS of UE traffic flows, and seamless handovers during UEs' mobility. In order to set up these use cases, we mainly discuss how to configure and evaluate ns-3 supporting LTE radio resource scheduling and handover algorithms.

In the first section of the chapter, we introduce how the ns-3 LTE module is modeling the operating spectrum in terms of frames, and **Resource Blocks (RBs)**. Then, we discuss the basic working of any ns-3-supporting LTE schedulers and how to configure various schedulers in simulations. As part of the research hands-on activity, we set up an LTE mobile network and evaluate the performance of ns-3 supporting LTE schedulers such as **Round Robin (RR)**, **Proportional Fair (PF)**, and **Channel and QoS aware (CQA)** schedulers.

In the second section of the chapter, we discuss how to configure the ns-3 LTE module supporting LTE EPS bearers for simulating QoS use cases with the help of the ns-3 LTE **Priority Set Scheduler (PSS)**. We discuss a hands-on activity related to setting up and evaluating QoS flow simulation in LTE networks. As part of this activity, we discuss simulation results in terms of **Guaranteed Bit Rate (GBR)** and non-GBR flows.

In the last section of the chapter, we start by introducing how LTE handover handling is implemented in ns-3 in terms of UE handover measurement configuration, various handover events' handling, and the importance of X2 interfaces. Then, we discuss ns-3 supporting handover algorithms and how to configure them in ns-3 LTE simulations. As part of the research hands-on activity, we set up an LTE mobility use case and evaluate ns-3 supporting handover algorithms.

We will discuss the following topics in this chapter:

- Exploring LTE radio resource scheduling algorithms supported in ns-3

- Simulating and testing LTE **Evolved Packet Core** (**EPC**) QoS bearers' supporting features in ns-3

- Setting up an LTE network mobility scenario using ns-3-supported handover algorithms

# Exploring LTE radio resource-scheduling algorithms supported in ns-3

One of the major challenges for LTE operators is the optimal allocation of a limited and costly operating spectrum (radio resources) to their users. Operators must configure radio resource schedulers on **evolved NodeBs** (**eNBs**, or cells) for allocating spectrum resources efficiently to their connected UEs. Let's understand how ns-3 models the LTE operating spectrum in terms of time and frequency chunks (radio resources) to support a variety of radio resource schedulers' implementation. In general, LTE networks can be deployed in **Frequency Division Duplex** (**FDD**) or **Time Division Duplex** (**TDD**) operating modes, but ns-3 supports only the LTE FDD operating mode for spectrum management. In LTE FDD mode, for uplink and downlink transmissions, separate bandwidth is allocated. Let's discuss ns-3 LTE operating spectrum modeling details in the following section.

## ns-3 LTE operating spectrum

As per the LTE standard implementation, the ns-3 LTE module models the LTE operating spectrum in terms of the following two dimensions: time and frequency. That is, spectrum resources are allocated to users in terms of time allocation units (such as **milliseconds**, or **ms**) and frequency resource units (such as kHz).

ns-3 models the LTE operating spectrum in terms of frequency resources as follows:

- As per the LTE standard, LTE networks can be deployed in the following operating bandwidths: 1.4 MHz, 3 MHz, 5 MHz, 10 MHz, 15 MHz, and 20 MHz.

- The minimum spectrum resource allocation unit is 180 kHz for any operating bandwidth. This is known as an RB. ns-3 LTE schedulers allocate spectrum resources to UEs in terms of RBs only.

- Moreover, as per the LTE standard, ns-3 LTE also ensures the following RBs and the corresponding operating bandwidths:

  1.4 MHz: 6 RBs; 3 MHz: 15 RBs; 5 MHz: 25 RBs; 10 MHz: 50 RBs; 15 MHz: 75 RBs; and 20 MHz: 100 RBs.

- ns-3 LTE simulation supports a minimum of 6 RBs and a maximum of 100 RBs.

- Moreover, ns-3 LTE supports FDD only. In simulations, it is necessary to determine and configure UL and DL bandwidths in terms of RBs.

ns-3 models the LTE operating spectrum in terms of time dimension resource allocation by considering the following:

- **Slot**: This is 0.5 ms in duration. A slot is the minimum time resource unit for which the LTE network can assign spectrum resources (RBs) to their users (or UEs). However, ns-3 LTE schedulers assign spectrum resources to various UEs for every 1 ms only.

- **Subframe**: This is 1 ms in duration. Two consecutive slots are considered as one subframe. ns-3 LTE schedulers assign spectrum resources to UEs at the subframe level only (1 ms).

- **Frame**: This is 10 ms in duration. 10 consecutive subframes are considered 1 frame. All major LTE operations such as broadcasting cell configuration, cell selection, random access, and scheduling details are communicated in the LTE network in terms of a frame or multiple frames.

Next, we discuss ns-3 LTE radio resource schedulers' design details.

## Common design considerations of ns-3 LTE radio resource schedulers

In ns-3 LTE simulations, radio resource schedulers run at the MAC layer of eNBs. For every **Time To Interval** (**TTI**) of 1 ms duration, schedulers running at eNBs allocate the necessary number of RBs to connected UEs' downlink and/or uplink transmissions. In order to optimally allocate RBs to UEs in terms of maximizing spectrum utilization and satisfying UE traffic requirements, there are many schedulers that evolved from research and are used in real-time deployments. ns-3 LTE supports a variety of scheduling algorithms:

- RR

- PF

- PSS

- CQA

- **Maximum Throughput (MT)**

- **Throughput to Average (TTA)**

- **Blind Average Throughput (BAT)**

- **Token Bank Fair Queue (TBFQ)**

Let's see the common implementation details of ns-3 LTE radio resource schedulers:

- Schedulers run at eNBs. For every TTI, based on the scheduling policy, RBs are allocated to UEs dynamically.

- UL and DL scheduling implementations are separated.

- Scheduling decisions are reported to UEs using the **Physical Downlink Control Channel** (**PDCCH**).

- The scheduling decision involves UE inputs such as channel conditions, traffic priority, buffer status, battery level, interference, and so on.

- RR schedulers divide all available RBs equally and assign them to UEs' flows in a round-robin manner. Hence, they do not ensure throughput fairness and GBR requirements for UEs.

- PF schedulers carefully select UEs and assign the necessary RBs based on their channel conditions and try to ensure throughput fairness among UEs' traffic flows. However, they do not guarantee any bit rates.

- In the case of EPS bearers being configured, the PSS scheduler handles the QoS requirements of radio bearers (in terms of GBRs). It mainly tries to ensure GBR requirements for UEs' traffic flows. However, it does not guarantee throughput fairness for UEs.

---

**Important note**

LTE schedulers' implementation is not specified by the **3rd Generation Partnership Project** (**3GPP**). Researchers/developers can check existing schedulers' code in the ns-3 LTE module and implement their own novel scheduling algorithms.

---

For example, under `src/lte/model`, we can check the `pf-ff-mac-scheduler.cc` and `pf-ff-mac-scheduler.h` files to explore the PF scheduler implementation details. Users can explore the `SchedDlTriggerReq` functions to understand the PF DL scheduler implementation and `SchedUlTriggerReq` to understand the PF UL scheduler implementation details.

Now, let's work through a hands-on activity to study a few of the ns-3-supporting LTE schedulers and their performance in the next section.

## ns-3 LTE schedulers' performance evaluation hands-on activity

In order to set up ns-3 LTE schedulers' performance evaluation in `pkt-lte-scheduler.cc`, we will do the following tasks:

- We extend `pkt-lte-epc.cc` (discussed in *Chapter 7*) to implement `pkt-lte-scheduler.cc` for conducting this hands-on activity.

- After `PathLossModel` configuration lines of code, configure `DlBandwidth` and `UlBandwidth` to 25 RBs for schedulers' evaluation.

- After RBs' configuration, configure ns-3 schedulers with RR, PF, and CQA for performance evaluation.

- We also configure the PSS scheduler for EPS bearers' evaluation. (However, this will be discussed in the LTE EPC QoS bearers section.)

- To make the performance evaluation discussion simple, we configure only DL application flows for UEs at the speed of 430 Kbps. Similarly, users can configure UL traffic flows.

- Evaluate schedulers' performance in the following two scenarios: 1) when UEs are experiencing good signal strength; 2) when UEs are suffering from interference due to neighbor eNBs.

Let's start doing some changes in pkt-lte-epc.cc to implement pkt-lte-scheduler.cc.

Start with creating UE deployment scenarios to simulate good signal strength in (scenario==1) and deploy some UEs near to eNBs, and for simulating an interference scenario in (scenario==2), deploy UEs in overlapping regions of eNBs using the following code snippet:

```
int scenario = atoi(argv[2]);
for (uint16_t i = 0; i < nUEs/2; i++)
  {
   if (scenario==1)
     positionAlloc->Add (Vector(i+10, 0, 0));
   if (scenario==2)
     positionAlloc->Add (Vector(i+100, 0, 0));
  }
for (uint16_t i = nUEs/2; i < nUEs; i++)
  {
   if (scenario==1)
     positionAlloc->Add (Vector(i+500, 0, 0));
   if (scenario==2)
     positionAlloc->Add (Vector(i+400, 0, 0));
  }
```

Another important change we need to do in pkt-lte-scheduler.cc is after the RBs' configuration line of code. Configure ns-3 LTE schedulers (RR, PF, CQA, and PSS) for performance evaluation using the following code snippet:

```
int sched = atoi(argv[1]);
int enablePSS = 0;
switch(sched)
{
  case 1:
    lteHelper->SetSchedulerType ("ns3::RrFfMacScheduler");
    break;
  case 2:
```

```
        lteHelper->SetSchedulerType ("ns3::PfFfMacScheduler");
        break;
    case 3:
        lteHelper->SetSchedulerType ("ns3::CqaFfMacScheduler");
        break;
    case 4:
        lteHelper->SetSchedulerType ("ns3::PssFfMacScheduler");
        lteHelper->SetSchedulerAttribute("nMux",
UintegerValue(15));
        lteHelper->SetSchedulerAttribute("PssFdSchedulerType",
StringValue("CoItA"));
        enablePSS = 1;
        break;
    }
```

That's all we need to do to evaluate the ns-3 schedulers' performance. In the next section, we will be evaluating ns-3 LTE schedulers' performance using `pkt-lte-scheduler.cc` thoroughly.

## ns-3 LTE schedulers' performance evaluation

We start with the evaluation of schedulers' performance in the scenario of UEs experiencing good signal strengths. In order to select a particular scheduler and evaluate its performance, we can pass the following arguments as command-line arguments: 1 for the RR scheduler, 2 for the PF scheduler, and 3 for the CQA scheduler.

Let's start evaluating RR scheduler performance in the scenario of UEs experiencing good signal strength. We can enable scenario 1 by passing 1 as a command-line argument using the following commands:

```
./ns3 run scratch/pkt-lte-scheduler -- "1" "1"
python3 scratch/flowmon-parse-results.py lte_sched.xml > rr_
sched
```

After executing these commands successfully, we have collected all UEs' throughput results into the `rr_sched` file.

Next, evaluate the PF scheduler performance using the following commands:

```
./ns3 run scratch/pkt-lte-scheduler -- "2" "1"
python3 scratch/flowmon-parse-results.py lte_sched.xml > pf_
sched
```

At end of the simulation, we have collected all UEs' throughput results into the `pf_sched` file.

Finally, evaluate the CQA scheduler performance using the following commands:

```
./ns3 run scratch/pkt-lte-scheduler -- "3" "1"
python3 scratch/flowmon-parse-results.py lte_sched.xml > cqa_
sched
```

At end of the simulation, we have collected all UEs' throughput results into the `cqa_sched` file.

Now, combine all scheduler UEs' throughput results into a single `comparesched1.csv` file to generate a chart using the following commands:

```
cat rr_sched |grep 'RX' |tail -n 40 | cut -f3 -d ' ' >rr.csv
cat pf_sched |grep 'RX' |tail -n 40 | cut -f3 -d ' ' >pf.csv
cat cqa_sched |grep 'RX' |tail -n 40 | cut -f3 -d ' ' >cqa.csv
paste rr.csv pf.csv cqa.csv > comparesched1.csv
```

Let's generate a line points chart for evaluating ns-3 LTE schedulers' performance using `gnuplot` or any plotting tool. We generated the following chart using **LibreOffice**:

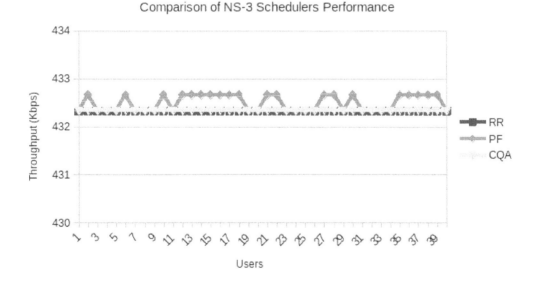

Figure 8.1 – Scenario 1: UEs are experiencing good signal strengths

*Figure 8.1* describes the ns-3 LTE radio resource schedulers' performance when all UEs are deployed near their connected eNBs. It shows the results of individual UEs' throughput. In *Figure 8.1*, blue (RR), red (PF), and yellow (CQA) lines describe the `RoundRobin`, `ProportionalFair`, and `ChannelandQoSAware` schedulers' performances. In the scenario of all UEs experiencing good signal strengths, their throughputs are not affected severely. Every UE is able to receive as per their DL application data rate (430 Kbps). Moreover, all schedulers are providing fair and better throughput for all UEs. Similar to the UEs' throughput plot, we can also plot delay and packet losses.

In the case of UEs deployed near their connected eNBs (cells), then the RR, PF, and CQA schedulers perform similarly. But in a real-time scenario, it is common for UEs to experience poor signal strengths and interference in the network.

Let's start evaluating RR scheduler performance in the case of UEs experiencing interference from neighbor eNBs using the following commands at your ns-3 installation directory. We can enable the interference scenario by passing 2 as the command-line argument using the following commands:

```
./ns3 run scratch/pkt-lte-scheduler -- "1" "2"
python3 scratch/flowmon-parse-results.py lte_sched.xml > rr_
sched
```

At end of the simulation, we have collected all UEs' throughput results into the rr_sched file.

Next, evaluate the PF scheduler performance using the following commands:

```
./ns3 run scratch/pkt-lte-scheduler -- "2" "2"
python3 scratch/flowmon-parse-results.py lte_sched.xml > pf_
sched
```

At end of the simulation, we have collected all UEs' throughput results into the pf_sched file.

Finally, evaluate the CQA scheduler performance using the following commands:

```
./ns3 run scratch/pkt-lte-scheduler -- "3" "2"
python3 scratch/flowmon-parse-results.py lte_sched.xml > cqa_
sched
```

At end of the simulation, we have collected all UEs' throughput results into the cqa_sched file.

Now, combine all schedulers' UEs throughput results into a single comparesched2.csv file for generating a chart using the following commands:

```
cat rr_sched |grep 'RX' |tail -n 40 | cut -f3 -d ' ' >rr.csv
cat pf_sched |grep 'RX' |tail -n 40 | cut -f3 -d ' ' >pf.csv
cat cqa_sched |grep 'RX' |tail -n 40 | cut -f3 -d ' ' >cqa.csv
paste rr.csv pf.csv cqa.csv > comparesched2.csv
```

Let's generate a line point chart to evaluate the ns-3 LTE schedulers' performance using `gnuplot` or any plotting tool. We generated the following chart using **LibreOffice**:

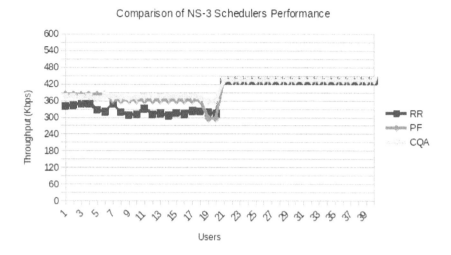

Figure 8.2 – Scenario 2: eNB-1's UEs are experiencing interference

In our simulation setup, UEs of eNB-1 are experiencing poor signal strengths and interference because UEs of eNB-1 are deployed in the overlapping region of eNB-1 and eNB-2 in our simulation setup. Hence, we focus on discussing the results of eNB-1 UEs in more detail. We have plotted the individual UEs' throughput results in *Figure 8.2*. From *Figure 8.2*, we can observe that PF and CQA schedulers are performing similarly in terms of offering throughput to UEs irrespective of their signal strengths and interference conditions. Since eNB-1's UEs are experiencing poor signal strengths and high interference, they are experiencing lower throughput (around 380 Kbps). On the other hand, eNB-2's UEs are experiencing higher throughput (around 420 Kbps) as per their DL applications' configured data rate, but PF and CQA schedulers are ensuring fairness for all UEs in a cell (or eNB) in terms of throughput. The reason for PF and CQA schedulers' better performance is due to consideration of UEs' signal strengths and their past and current throughput to make their scheduling decisions. You may have doubts about this because CQA is not performing in a special way compared to PF. This is due to a lack of EPS bearers (GBR) configured for UE traffic flows. We discuss the simulation and evaluation of EPS bearers and QoS scenarios in the next section.

On other hand, the RR scheduler is performing poorly compared to the PF and CQA schedulers. This is due to its RB allocation policy. The RR scheduler blindly allocates an equal number of RBs to UE flows irrespective of UE signal strengths, past and current throughput, and EPS bearers. Hence, RR scheduling decisions are leading to poor performance in terms of ensuring UE throughput.

Till now, we have discussed UE throughput results only. It is also possible to plot delay and packet loss ratio results from the LTE flow metrics files. Let's see how to create a delay plot to analyze the schedulers' performance using the following commands:

```
cat rr_sched |grep 'Mean' |tail -n 40 | cut -f3 -d ' '
>rrdelay.csv
cat pf_sched |grep 'Mean' |tail -n 40 | cut -f3 -d ' '
>pfdelay.csv
cat cqa_sched |grep 'Mean' |tail -n 40 | cut -f3 -d ' '
>cqadelay.csv
paste rrdelay.csv pfdelay.csv cqadelay.csv > delaysched.csv
```

Now, let's generate a delay plot using the delaysched.csv file. We generated the following chart using **LibreOffice**:

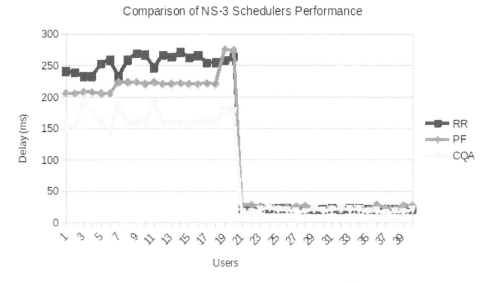

Figure 8.3 – Individual UEs experiencing latencies in an interference scenario

The results in *Figure 8.3* show that when UE channel conditions are poor, UEs face higher delays and variations. Moreover, deploying simple schedulers such as RR leads to higher delays for UEs (shown with the blue line). On the other hand, having intelligent schedulers such as PF and CQA on the eNBs provides lower delays and fairness among UEs.

Similarly, we can plot the packet loss ratio results using the following commands:

```
cat rr_sched |grep 'Packet' |tail -n 40 | cut -f4 -d ' '
>rrplr.csv
cat pf_sched |grep 'Packet' |tail -n 40 | cut -f4 -d ' '
>pfplr.csv
cat cqa_sched |grep 'Packet' |tail -n 40 | cut -f4 -d ' '
>cqaplr.csv
paste rrplr.csv pfplr.csv cqaplr.csv > plrcomp.csv
```

Now, let's generate a packet loss ratio plot using the `plrcomp.csv` file. We generated the following chart using **LibreOffice**:

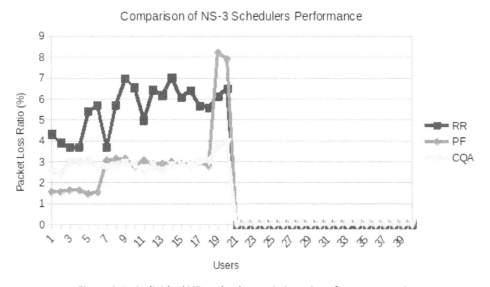

Figure 8.4 – Individual UE packet loss ratio in an interference scenario

When UE channel conditions are poor and we deploy simple schedulers such as RR, especially at eNB-1, it leads to higher packet loss ratios for UEs (shown in *Figure 8.4* with a blue line). On the other hand, deploying intelligent schedulers at eNB-1, such as PF and CQA, minimizes packet losses to UE flows.

Well done! We have successfully evaluated a few of the ns-3 LTE schedulers' performance using a hands-on activity. Similarly, users can explore other schedulers by configuring them in their own time. Next, we discuss how to simulate ns-3-supporting LTE QoS bearers and testing using a hands-on activity.

# Simulating and testing LTE EPC QoS bearers' supporting features in ns-3

One of the interesting features of LTE networks is ensuring QoS requirements for UEs' traffic flows. As we discussed in the previous chapter (*Chapter 7*), to meet QoS requirements of traffic flows, LTE networks establish end-to-end EPS bearers from UEs to P-GW. Based on the QoS requirements of a flow, a unique EPS bearer will be established and is mapped to sub-bearers: UE and eNB (radio bearer), eNB and S-GW (s1 bearer), S-GW and P-GW, (s5 bearer). Radio bearers are helpful at the LTE-RAN level for allocating necessary radio resources to meet the requirements of a QoS flow. Besides, s1 and s5 bearers help to ensure necessary bandwidth requirements at the EPC level. However, the ns-3.36 LTE module supports only the simulation of GBR and non-GBR flows. Moreover, it is assumed that abundant link capacity is available at EPC.

The ns-3 EPS bearers' simulation has the following limitations:

- No **allocation and retention priority** is implemented to decide whether particular EPC bearers can be accepted or rejected
- No provision to guarantee the latency requirements of flows
- Users must configure the PSS scheduler to test GBR flows

In order to simulate GBR flows in ns-3 LTE simulations, we must configure the PSS scheduler and set up the necessary EPS bearers. Let's simulate and test the LTE EPC QoS bearer using the following hands-on activity. In order to implement the hands-on activity, do the following tasks:

- Extend the `pkt-lte-epc.cc` simulation to simulate GBR flows.
- Deploy eNBs and their respective UEs with poor signal strength in your simulation environment.
- Configure the PSS scheduler.
- Configure the LTE operating bandwidth as 25 RBs.
- Configure single GBR bit rate (around 430 Kbps) flows for all eNB-1-connected UEs.
- Configure two-level GBR flows for eNB-2 connected UEs. Configure the GBR level-1 maximum bit rate to around 430 Kbps for the first 10 UEs, and the GBR level-2 maximum bit rate to around 128 Kbps for another 10 UEs.
- Configure only DL UDP flows with a data rate of 430 Kbps for simulation.
- Plot single-level GBR flows and two-level GBR flow results for evaluation and inspection.

Let's discuss the necessary changes that should be made in `pkt-lte-scheduler.cc` to simulate EPS QoS bearers.

Start with eNB and UE deployment changes, as per the hands-on activity:

```
for (uint16_t i = 0; i < nUEs/2; i++)
  {
    positionAlloc->Add (Vector(i+100, 0, 0));
  }
for (uint16_t i = nUEs/2; i < nUEs; i++)
  {
    positionAlloc->Add (Vector(i+400, 0, 0));
  }
```

Next, configure the PSS scheduler to ensure the GBR requirements of traffic flows:

```
lteHelper->SetSchedulerType ("ns3::PssFfMacScheduler");
lteHelper->SetSchedulerAttribute("nMux", UintegerValue(15));
lteHelper->SetSchedulerAttribute ("PssFdSchedulerType",
StringValue("CoItA"));
```

Finally, before the UDP applications' configuration lines of code, configure EPS bearers for UEs (1 to 30) DL flows with GBR level-1 (430 Kbps), and for UEs (31 to 40) with GBR level-2 bit rates (128 Kbps) using the following code snippet:

```
for (uint32_t u = 0; u <nUEs; u++)
  {
    Ptr<NetDevice> ueDevice = UEdevs.Get (u);
    GbrQosInformation qos;
      if (u<30)
{
      qos.gbrDl = (512+32)*100*8;
      }
      else
      {
        qos.gbrDl = (128+32)*100*8;
      }
    qos.mbrDl = qos.gbrDl;
```

```
        enum EpsBearer::Qci q = EpsBearer::GBR_CONV_VOICE;
        EpsBearer bearer (q, qos);
        lteHelper->ActivateDedicatedEpsBearer (ueDevice, bearer,
EpcTft::Default ());
    }
```

After saving all these changes in `pkt-lte-scheduler.cc`, let's evaluate it in the next section.

## Evaluation of LTE EPS QoS bearers

Let's first evaluate our simulation setup with a single GBR (around 430 Kbps) for all UEs' (1 to 40) flows (change the respective UE bearers' configuration code in `pkt-lte-scheduler.cc`) using the following commands:

```
./ns3 run scratch/pkt-lte-scheduler -- "4" "2"
python3 scratch/flowmon-parse-results.py lte_sched.xml > pss_
sched1
cat pss_sched1 |grep 'RX' |tail -n 40 | cut -f3 -d ' ' >pss1.
csv
```

Here, we collected UE flows' throughput results into the `pss1.csv` file.

Next, to check the working of EPS bearers according to GBR levels, let's change `pkt-lte-scheduler.cc` to configure bit rates of around 430 Kbps for the first 30 UEs and 128 Kbps for the remaining UEs. Then, execute the following commands:

```
./ns3 run scratch/pkt-lte-scheduler --  "4" "2"
python3 scratch/flowmon-parse-results.py lte_sched.xml > pss_
sched2
cat pss_sched2 |grep 'RX' |tail -n 40 | cut -f3 -d ' ' >pss2.
csv
```

Here, we collected UE flows' throughput results into the `pss2.csv` file.

Finally, collect the `pss1.csv` and `pss2.csv` results into `psscomp.csv` for plotting results:

```
paste pss1.csv pss2.csv > psscomp.csv
```

We used `psscomp.csv` and generated a line points chart (shown in *Figure 8.5*) using **LibreOffice**:

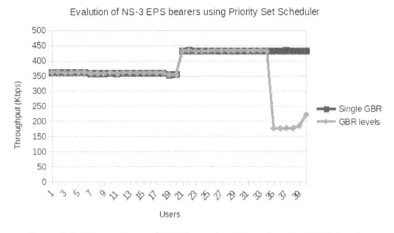

Figure 8.5 – Comparison of UEs' throughput based on their GBR levels

In *Figure 8.5*, the blue line (**Single GBR**) describes the results in the case of a single GBR for all UEs at the eNB. It is showing throughput of eNB-1's and eNB-2's UEs. As all UEs (1 to 20) of eNB-1 and UEs (21 to 40) of eNB-2 are configured with the same GBR, UEs experience the same throughput from their eNB. This means as per the EPS bearers configured in our simulation, the ns-3 LTE module ensures the bit rates for UEs with the help of the PSS scheduler at the eNBs.

On the other hand, the red line (**GBR levels**) describes the results in the case of two GBR levels for UEs of eNB-2. We simulate this scenario to explore the working of multiple parallel EPS bearers. In our simulation, UEs (21 to 30) of eNB-2 are configured with a 430 Kbps GBR level, hence they experience better throughput compared to other UEs. Especially, UEs (35 to 40) experience around 150 Kbps only due to their GBR level (128 Kbps) configuration. That means as per the EPS bearers, the PSS scheduler tries to ensure the bit rates of UEs. Hence, we can see clearly that the LTE EPS bearers and PSS are highly useful for simulating LTE network QoS scenarios. Next, let's set up LTE mobility scenario simulations to evaluate the ns-3 supporting LTE handover algorithms.

# Setting up an LTE network mobility scenario using ns-3-supported handover algorithms

In LTE networks, one of the important promising features is the support of seamless mobility for mobile users. In order to support seamless mobility, the LTE standard defines various handover measurement reports, events, and signaling messages. To support simulations to evaluate LTE mobility scenarios, the ns-3 LTE module supports the following features:

- Generation and handling of handover measurement reports
- LTE standard handover events' simulation
- Handover algorithms

# Generation and handling of handover measurement reports

As we discussed in *Chapter 7*, ns-3 LTE UEs generate RSRQ and RSRP handover measurement reports for supporting handovers. At a high level, RSRQ reports indicate SINR values perceived by UEs in the network from the connected eNB or neighbor eNBs, whereas RSRP reports indicate the signal strength received from the connected eNB or neighbor eNBs. In standard LTE networks, eNBs have to configure their connected UEs to receive handover measurement reports. Similarly, the ns-3 LTE module also supports eNBs to configure their UEs to get handover measurement reports. By receiving handover measurement reports from UEs, eNBs can make informed UE-handover-related decisions.

# LTE standard handover events' simulation

Usually, many UEs get served by one eNB, hence processing every RSRQ and RSRP measurement report from UEs at the eNB will be a heavy load. To limit and process only necessary reports, the LTE standard supports the following handover measurement events to make handover decisions. This mainly helps eNBs to reduce unnecessary signaling overhead and optimizes handover-related spectrum resources. It also allows researchers to propose and implement novel and optimal handover algorithms. Our ns-3 LTE module also supports LTE handover events for implementing various handover algorithms. ns-3 supports the following UE-triggered handover events. These events are triggered by UEs based on handover measurement reports' generation configuration details provided by eNB:

- **Event A1**: UE generates Event A1 to the connected eNB when it is perceiving the RSRQ from the connected eNB (serving cell) as becoming better than a threshold.

- **Event A2**: UE generates Event A2 to the connected eNB when it is perceiving the RSRQ from the connected eNB (serving cell) as becoming worse than a threshold.

- **Event A3**: UE generates Event A3 to the connected eNB when it is perceiving the RSRP from a neighbor cell offset as becoming better than the RSRP from the connected eNB (serving cell).

- **Event A4**: UE generates Event A4 to the connected eNB when it is perceiving the RSRQ from a neighbor cell as becoming better than a threshold.

- **Event A5**: UE generates Event A5 to the connected eNB when it is perceiving the RSRQ from the connected eNB (serving cell) as becoming worse than `threshold1` and the RSRQ from a neighbor cell as becoming better than `threshold2`.

LTE engineers/researchers can use these events for implementing novel handover algorithms. Next, let's discuss the ns-3 LTE module supporting handover algorithms.

## ns-3.36 LTE module supporting handover algorithms

The ns-3 LTE module supports the following handover algorithms:

- `A2A4RsrqHandoverAlgorithm`
- `A3RsrpHandoverAlgorithm`

**A2-A4-RSRQ handover algorithm:** This is implemented using LTE handover events A2 and A4. Hence, eNBs running `ns3::A2A4RsrqHandoverAlgorithm` must configure `ServingCellThreshold` and `NeighbourCellOffset` thresholds (dB) for receiving handover measurement reports from UEs. Hence, UEs generate A2 and A4 events when the RSRQ from the connected eNB (serving cell) becomes worse than the `ServingCellThreshold` threshold and the neighbor cell's RSRQ from the serving cell RSRQ becomes better than the `NeighbourCellOffset` threshold.

By receiving A2 and A4 events from UEs, `A2A4RsrqHandoverAlgorithm` running on the eNB decides whether to hand over its UEs to the neighbor eNBs.

Let's see how to configure `ns3::A2A4RsrqHandoverAlgorithm` with the necessary threshold values in our mobility simulations using the following code snippet:

```
lteHelper->SetHandoverAlgorithmType
("ns3::A2A4RsrqHandoverAlgorithm");
  lteHelper->SetHandoverAlgorithmAttribute
("ServingCellThreshold",

                                      UintegerValue (30));
  lteHelper->SetHandoverAlgorithmAttribute
("NeighbourCellOffset",

                                      UintegerValue (5));
```

**A3-RSRP-Handover algorithm:** This is also known as the strongest cell handover algorithm. It is implemented using Event A3. In the case of using `ns3::A3RsrpHandoverAlgorithm` on eNBs, their connected UEs must be configured with `Hysteresis` (dB) and `TimeToTrigger` (ms) attributes to generate A3 events. Hence, UEs generate A3 events to connected eNBs.

On receiving A3 events from UEs, `A3RsrpHandoverAlgorithm` running on the eNBs checks if the difference between the serving cell RSRP and the neighbor cell RSRP is greater than the `Hysteresis` threshold for the duration of `TimeToTrigger`; only then does it decide to hand over the UE to neighbor eNBs. Here, `Hysteresis` and `TimeToTrigger` are useful to avoid UE handover oscillations (also known as ping-pong handovers). Let's see how to configure `A3RsrpHandoverAlgorithm` with the necessary threshold values in our mobility scenarios using the following code snippet:

```
lteHelper->SetHandoverAlgorithmType
("ns3::A3RsrpHandoverAlgorithm");
lteHelper->SetHandoverAlgorithmAttribute ("Hysteresis",
```

```
                                                              DoubleValue
(3.0));
lteHelper->SetHandoverAlgorithmAttribute ("TimeToTrigger",
                                                              TimeValue
(MilliSeconds (256)));
```

> **Important note**
>
> List of allowed `TimeToTrigger` configurable values: 0, 40, 64, 80, 100, 128, 160, 256, 320, 480, 512, 640, 1024, 1280, 2560, and 5120.
>
> Default values for `A3RsrpHandoveAlgorithm`: `Hysteresis`: 3 dB; `TimeToTrigger`: 256 ms.
>
> Default values of `A2A4RsrqHandoverAlgorithm`: `NeighborCellOffset`: 1 dB; `ServingCellThreshold`: 30 dB

Researchers and engineers can propose and implement novel handover algorithms by writing a subclass of the `LteHandoverAlgorithm` abstract superclass.

Before going on to implement a hands-on activity related to LTE mobility simulations, note the following important points:

- It is necessary to connect ns-3 eNBs using the X2 interface to simulate LTE mobility scenarios
- Source eNB and target eNBs use the X2 interface for exchanging buffered data, sequence numbers, and handover signaling messages
- In ns-3 LTE simulations, it is also possible to trigger manual handover for UEs
- It is necessary to configure ns-3 supporting `A2A4RsrqHandoverAlgorithm` or `A3RsrpHandoverAlgorithm` to simulate automatic handovers in ns-3 LTE network simulations

That's excellent. Let's start implementing an LTE mobility scenario simulation using ns-3 in the next section.

## ns-3 LTE mobility scenario simulation

In this activity, we mainly focus on how to quickly set up LTE mobility scenarios and evaluate ns-3 supporting handover algorithms. In order to simulate and evaluate an LTE mobility scenario, we do the following tasks using the ns-3 simulation topology (shown in *Figure 8.6*) in the hands-on activity:

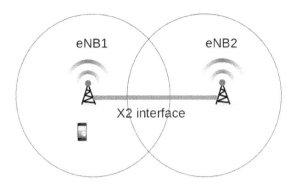

Figure 8.6 – LTE mobility scenario simulation topology

- Two eNBs are deployed as shown in *Figure 8.6*
- 20 UEs are deployed near eNB-1 and another 20 UEs are deployed near eNB-2
- All UEs are static, except UE-18 and UE-21
- UE-18 is initially attached to eNB-1 and UE-21 is initially attached to eNB-2
- UE-18 is moving toward eNB-2 at a speed of 5 ms, and UE-21 is moving toward eNB-1 at a speed of 10 ms
- During the simulation, we want to track UEs' locations
- Simulate handovers using the `A2A4RsrqHandover` algorithm
- Simulate handovers using `A3RsrpHandover` algorithms
- During the evaluation of handover algorithms, track at what time UEs are handed over to neighbor eNBs, and check handover delays
- Evaluate handover algorithms using the default thresholds and by changing the default values

Let's start implementing our LTE mobility scenario in `pkt-lte-handovers.cc` by importing the following packages:

```
#include "ns3/lte-helper.h"
#include "ns3/epc-helper.h"
#include "ns3/core-module.h"
#include "ns3/network-module.h"
#include "ns3/ipv4-global-routing-helper.h"
#include "ns3/internet-module.h"
#include "ns3/mobility-module.h"
#include "ns3/lte-module.h"
```

```
#include "ns3/applications-module.h"
#include "ns3/point-to-point-helper.h"
#include "ns3/config-store.h"
#include "ns3/flow-monitor-module.h"
using namespace ns3;
```

Before starting the main() implementation, write the following functions to trace the handover start and end events from the eNBs. The following are the handover events tracing callback functions, which will be linked in main():

```
void
EnbStartsHandover (std::string context,
                        uint64_t imsi,
                        uint16_t cellid,
                        uint16_t rnti,
                        uint16_t targetCellId)
{
  std::cout << context << "Handover Start time:" <<
Simulator::Now ().GetSeconds ();
  std::cout
            << " eNB CellId " << cellid
            << ": start handover of UE with IMSI " << imsi
            << " RNTI " << rnti
            << " to CellId " << targetCellId<<"
Time:"<<Simulator::Now()
            << std::endl;
}
void
EnbHandoverEnd (std::string context,
                        uint64_t imsi,
                        uint16_t cellid,
                        uint16_t rnti)
{
  std::cout << context << "Handover End time:" <<
Simulator::Now ().GetSeconds ();
  std::cout
            << " eNB CellId " << cellid
```

```
          << ": completed handover of UE with IMSI " << imsi
          << " RNTI " << rnti <<" Time:"<<Simulator::Now()
          << std::endl;
}
```

As we want to simulate particular mobility patterns for UEs in this activity, the following functions are implemented. We update the locations of the UEs in the simulation using the following function:

```
static void
SetLoc (Ptr<Node> ue, Vector loc)
{
  Ptr<MobilityModel> mobility = ue->GetObject<MobilityModel>
();
  mobility->SetPosition (loc);
}
```

We get the current location of a UE in the simulation using the following function:

```
static Vector
GetLoc (Ptr<Node> ue)
{
  Ptr<MobilityModel> mobility = ue->GetObject<MobilityModel>
();
  return mobility->GetPosition ();
}
```

We move the UE with a specific speed in the simulation using the following function:

```
static void
MoveUE (Ptr<Node> ue, int dist)
{
  Vector loc = GetLoc (ue);
  std::cout<<" UE location:"<<loc.x<<","<<loc.y<<"\n";
  loc.x = loc.x + dist;
  SetLoc (ue, loc);
}
```

After completing the handover events' tracing and mobility pattern functions' implementations, start `main()` by creating the necessary number of eNBs and UEs for our activity:

```
int
main (int argc, char *argv[])
{
  uint16_t nUEs = 40;
  double simTime = 30.0;
  NodeContainer UEs;
  NodeContainer eNBs;
  eNBs.Create(2);
  UEs.Create(nUEs);
```

Then, deploy eNBs at specific locations. After that, place the first 20 UEs near eNB-1 and the next 20 UEs near eNB-2 using the ns-3 placement and mobility models:

```
  Ptr<ListPositionAllocator> positionAlloc =
CreateObject<ListPositionAllocator> ();
  positionAlloc->Add (Vector(0, 0, 0));
  positionAlloc->Add (Vector(300, 0, 0));
  for (uint16_t i = 0; i < nUEs/2; i++)
    {
      positionAlloc->Add (Vector(i+100, 0, 0));
    }
  for (uint16_t i = nUEs/2; i < nUEs; i++)
    {
      positionAlloc->Add (Vector(300+i, 0, 0));
    }
  MobilityHelper mobility;
  mobility.
SetMobilityModel("ns3::ConstantPositionMobilityModel");
  mobility.SetPositionAllocator(positionAlloc);
  mobility.Install(eNBs);
  mobility.Install(UEs);
```

After setting up UEs' placement in our simulation topology, configure eNBs' and UEs' transmission power, enable data and control signals error modeling, and set SRS periodicity:

```
  Config::SetDefault ("ns3::LteEnbPhy::TxPower", DoubleValue
(46.0));
```

```
  Config::SetDefault ("ns3::LteUePhy::TxPower", DoubleValue
(20.0));
  Config::SetDefault
("ns3::LteSpectrumPhy::CtrlErrorModelEnabled",
BooleanValue(true));
  Config::SetDefault
("ns3::LteSpectrumPhy::DataErrorModelEnabled",
BooleanValue(true));
  Config::SetDefault ("ns3::LteEnbRrc::SrsPeriodicity",
UintegerValue (80));
```

Then, configure the path loss and propagation loss models and the UL and DL bandwidths of our LTE simulation networks:

```
  Ptr<LteHelper> lteHelper = CreateObject<LteHelper> ();
  Ptr<PointToPointEpcHelper>  epcHelper =
CreateObject<PointToPointEpcHelper> ();
  lteHelper->SetEpcHelper (epcHelper);
  lteHelper->SetAttribute ("PathlossModel", StringValue
("ns3::FriisPropagationLossModel"));
  lteHelper->SetEnbDeviceAttribute ("DlBandwidth",
UintegerValue (50));
  lteHelper->SetEnbDeviceAttribute ("UlBandwidth",
UintegerValue (50));
```

Next, enable the ns-3 supporting handover algorithms, and configure the A2A4RsrqHandoverAlgorithm parameters using the following code snippet:

```
  int algorithm = atoi(argv[1]);
  if (algorithm == 1)
  {
    lteHelper->SetHandoverAlgorithmType
("ns3::A2A4RsrqHandoverAlgorithm");
    lteHelper->SetHandoverAlgorithmAttribute
("ServingCellThreshold",
                                          UintegerValue
(30));
    lteHelper->SetHandoverAlgorithmAttribute
("NeighbourCellOffset",
                                          UintegerValue (5));
  }
```

Similarly, configure the `A3RsrpHandoverAlgorithm` parameters using the following code snippet:

```
if (algorithm == 2)
{
   lteHelper-
>SetHandoverAlgorithmType  ("ns3::A3RsrpHandoverAlgorithm");
   lteHelper->SetHandoverAlgorithmAttribute ("Hysteresis",
                                       DoubleValue
(2.0));
   lteHelper->SetHandoverAlgorithmAttribute ("TimeToTrigger",
                                       TimeValue
(MilliSeconds (100)));
}
```

Install the LTE protocols of eNBs and UEs using the following code snippet:

```
NetDeviceContainer eNBdevs = lteHelper->InstallEnbDevice
(eNBs);
NetDeviceContainer UEdevs = lteHelper->InstallUeDevice (UEs);
```

After setting up common LTE RAN configurations, to simulate internet services access from the LTE network, we must create an internet host and connect with P-GW using the P2P channel explicitly, as follows.

First set up an internet host to install our testing application using the following lines of code:

```
NodeContainer internetHost;
internetHost.Create (1);
InternetStackHelper internet;
internet.Install (internetHost);
```

Now connect the internet host with P-GW using the P2P channel of 10 Gbps speed using the following lines of code:

```
Ptr<Node> ihost = internetHost.Get (0);
Ptr<Node> pgw = epcHelper->GetPgwNode ();
PointToPointHelper p2p;
p2p.SetDeviceAttribute ("DataRate", DataRateValue (DataRate
("10Gb/s")));
p2p.SetDeviceAttribute ("Mtu", UintegerValue (1500));
p2p.SetChannelAttribute ("Delay", TimeValue (Seconds
(0.010)));
NetDeviceContainer inetDevs = p2p.Install (pgw,ihost);
```

In ns-3 LTE simulations, we must configure a 1.0.0.0/8 subnet for the P-GW and internet host interfaces. Then, it is also necessary to set up a static route on the internet host to reach the ns-3 LTE UEs, as follows:

```
Ipv4AddressHelper ip;
ip.SetBase ("1.0.0.0", "255.0.0.0");
Ipv4InterfaceContainer ipinterfs = ip.Assign (inetDevs);
Ipv4Address hostaddr = ipinterfs.GetAddress (1);
Ipv4StaticRoutingHelper iproute;
Ptr<Ipv4StaticRouting> ihostroute  = iproute.GetStaticRouting
(ihost->GetObject<Ipv4> ());
ihostroute->AddNetworkRouteTo (Ipv4Address ("7.0.0.0"),
Ipv4Mask ("255.0.0.0"), 1);
```

Besides, you must also configure all UE IP addresses using the ns-3 EpcHelper application and set up the default gateway address for all UEs explicitly using static routing:

```
internet.Install (UEs);
Ipv4InterfaceContainer ueIPs;
ueIPs = epcHelper->AssignUeIpv4Address (NetDeviceContainer
(UEdevs));
for (uint32_t i = 0; i < UEs.GetN (); ++i)
   {
     Ptr<Node> ue = UEs.Get (i);
     Ptr<Ipv4StaticRouting> ueroute = iproute.GetStaticRouting
(ue->GetObject<Ipv4> ());
     ueroute->SetDefaultRoute (epcHelper-
>GetUeDefaultGatewayAddress (), 1);
   }
```

Till this point, we have done all necessary ns-3 LTE RAN and EPC configurations to simulate internet service access.

Now, let's attach the LTE UEs to the LTE network manually. Specifically, connect half of the UEs to eNB-1 and the remaining UEs to eNB-2 using the following lines of code:

```
for (uint16_t i = 0; i < nUEs/2; i++)
   {
     lteHelper->Attach (UEdevs.Get(i), eNBdevs.Get(0));
   }
for (uint16_t i = nUEs/2; i < nUEs; i++)
   {
```

```
        lteHelper->Attach (UEdevs.Get(i), eNBdevs.Get(1));
    }
```

Next, start setting up DL flows (internet host -> UEs) for evaluation of our ns-3 LTE simulation setup.

Let's set up a unique DL flow between each UE and internet host using ns-3 `UdpClient` and ns-3 `PacketSink` applications. In the following loop, install UE-specific `PacketSink` applications on the UEs and respective `UdpClient` applications on the internet host. Here, we configure a UDP application data rate of around 430 Kbps for each DL flow (internet host to UE) using the following lines of code:

```
uint16_t dlPort = 5001;
ApplicationContainer clientApps;
ApplicationContainer serverApps;
for (uint32_t u = 0; u < UEs.GetN (); ++u)
  {
    PacketSinkHelper dlPacketSinkHelper
("ns3::UdpSocketFactory", InetSocketAddress
(Ipv4Address::GetAny (), dlPort));
    serverApps.Add (dlPacketSinkHelper.Install (UEs.Get(u)));
    UdpClientHelper dlClient (ueIPs.GetAddress (u), dlPort);
    dlClient.SetAttribute ("Interval", TimeValue
(Seconds(0.01)));
    dlClient.SetAttribute ("PacketSize", UintegerValue(512));
    dlClient.SetAttribute ("MaxPackets",
UintegerValue(10000));
    clientApps.Add (dlClient.Install (ihost));
    dlPort++;
  }
serverApps.Start (Seconds (1.01));
clientApps.Start (Seconds (1.01));
```

Next, for viewing various LTE simulation results, enable LTE protocol layers' level statistics collection:

```
lteHelper->EnableTraces ();
```

Connect all eNBs in our simulation using the X2 interface to simulate handovers:

```
lteHelper->AddX2Interface (eNBs);
```

Configure eNBs handovers' start and end event tracing to track UE handover events using the following lines of code:

```
Config::Connect ("/NodeList/*/DeviceList/*/LteEnbRrc/
HandoverStart",
                    MakeCallback (&EnbStartsHandover));
  Config::Connect ("/NodeList/*/DeviceList/*/LteEnbRrc/
HandoverEndOk",
                    MakeCallback (&EnbHandoverEnd));
```

To simulate our hands-on activity-specific mobility pattern for UE-18 and UE-21, we can use `Simulator::Schedule` to generate mobility events using the following code snippet:

```
int dist=5;
for(double t=1.0;t<simTime;t=t+1.0)
{
   Simulator::Schedule (Seconds (t),&MoveUE,UEs.Get(18),dist);
   Simulator::Schedule (Seconds (t),&MoveUE,UEs.Get(21),-
(dist+5));
}
```

Finally, close `main()` by setting up a flow monitor to collect flow level stats (such as throughput, delay, and packet loss) and free all simulation resources using the following lines of code:

```
Ptr<FlowMonitor> flowmon;
FlowMonitorHelper flowmonHelper;
flowmon = flowmonHelper.InstallAll ();
Simulator::Stop(Seconds(simTime));
Simulator::Run();
flowmon->SerializeToXmlFile ("lte_flowmetrics.xml", true,
true);
Simulator::Destroy();
return 0;
}
```

That's superb! We have set up our LTE network mobility scenarios for evaluating handover algorithms simulations successfully using ns-3. Without waiting, let's see how to evaluate it thoroughly in the next section.

## ns-3 LTE mobility scenario evaluation

In our LTE mobility scenario evaluation, we need to check the seamless handover of UEs, handover traces from eNBs, and handover delays. We check these details by evaluating the following ns-3 supporting handover algorithms:

- A2A4-RSRQ-Handovers
- A3-RSRP-Handovers

Let's start evaluating A2A4RsrqHandoverAlgorithm using the following commands with default ServingCellThreshold (30 dB) and NeighbourCellOffset (1 dB) values for executing pkt-lte-handovers.cc at the ns-3.36 installation directory. While executing the simulation, we can observe the following details:

```
./ns3 run scratch/pkt-lte-handovers - "1"
..
UE location:271,0
UE location:148,0
UE location:261,0
/NodeList/0/DeviceList/0/LteEnbRrc/HandoverStartHandover Start
time:7.52 eNB CellId 1: start handover of UE with IMSI 19 RNTI
16 to CellId 2 Time:+7.52e+09ns
/NodeList/1/DeviceList/0/LteEnbRrc/HandoverEndOkHandover End
time:7.52421 eNB CellId 2: completed handover of UE with IMSI
19 RNTI 29 Time:+7.52421e+09ns
UE location:153,0
UE location:251,0
..
..
UE location:203,0
..
UE location:151,0
/NodeList/1/DeviceList/0/LteEnbRrc/HandoverStartHandover Start
time:18.6 eNB CellId 2: start handover of UE with IMSI 22 RNTI
21 to CellId 1 Time:+1.86e+10ns
/NodeList/0/DeviceList/0/LteEnbRrc/HandoverEndOkHandover End
time:18.6042 eNB CellId 1: completed handover of UE with IMSI
22 RNTI 29 Time:+1.86042e+10ns
UE location:208,0
```

As we enabled handover start and end traces, we can track the following important details related to UE-18 and UE-21 from the handover traces:

- UE-18 (IMSI 19) started handing over from eNB-1 (CellID-1) to eNB-2 (CellID-2) at 7.52e+09ns and the handover completed at 7.52421e+09ns on eNB-2

- The handover delay for UE-18 is (7.52421e+09ns)-(7.52e+09ns)

- UE-21 (IMSI 22) started handing over from eNB-2 (CellID-2) to eNB-1 (CellID-1) at 1.86e+10ns and the handover completed at 1.86042e+10ns on eNB-1

- The handover delay for UE-21 is (1.86042e+10ns) –(1.86e+10ns)

- We can also observe UEs' locations (X, Y) during simulation execution

Now, let's change the ServingCellThreshold(30dB) and NeighbourCellOffset(5dB) values for executing pkt-lte-handovers.cc. While executing the simulation using the command shown next, we can observe the following details:

```
./ns3 run scratch/pkt-lte-handovers - "1"
..
UE location:168,0
UE location:221,0
/NodeList/0/DeviceList/0/LteEnbRrc/HandoverStartHandover Start
time:11.36 eNB CellId 1: start handover of UE with IMSI 19 RNTI
16 to CellId 2 Time:+1.136e+10ns
/NodeList/1/DeviceList/0/LteEnbRrc/HandoverEndOkHandover End
time:11.3642 eNB CellId 2: completed handover of UE with IMSI
19 RNTI 29 Time:+1.13642e+10ns
UE location:173,0
UE location:211,0
..
..
UE location:213,0
UE location:131,0
/NodeList/1/DeviceList/0/LteEnbRrc/HandoverStartHandover Start
time:20.28 eNB CellId 2: start handover of UE with IMSI 22 RNTI
21 to CellId 1 Time:+2.028e+10ns
/NodeList/0/DeviceList/0/LteEnbRrc/HandoverEndOkHandover End
time:20.2842 eNB CellId 1: completed handover of UE with IMSI
22 RNTI 29 Time:+2.02842e+10ns
UE location:218,0
UE location:121,0
```

From these results, we can track the following important details related to UE-18 and UE-21 from the handover traces:

- UE-18 (IMSI 19) started handing over from eNB-1 (CellID-1) to eNB-2 (CellID-2) at 1.136e+10ns and the handover completed at 1.13642e+10ns on eNB-2. Here, we can note that the handover start was delayed compared to the previous evaluation. This is due to an increase in `NeigborCellOffset` from 1 dB to 5 dB.

- The handover delay for UE-18 is (1.13642e+10ns)–(1.136e+10ns). Although the handover start time was delayed, there is no change in the handover delay.

- UE-21 (IMSI 22) started handing over from eNB-2 (CellID-2) to eNB-1 (CellID-1) at 2.028e+10ns and the handover completed at 2.02842e+10ns on eNB-1. The UE-21 handover started from eNB-2 was also delayed due to an increase in `NeighbourCellOffset`.

- The handover delay for UE-21 is (2.02842e+10ns)–(2.028e+10ns). Although the handover start time was delayed, there is no change in the handover delay.

To conclude, the changes in `ServingCellThreshold` and `NeighborCellOffset` led to a delay in the handover start from the eNBs. Sometimes it may be helpful to UEs get served with less loaded eNBs, and on the other hand, it may lead to delay and packet losses. Hence, it is necessary to configure suitable handover thresholds to optimize handover performance.

These are good observations about `A2A4RsrqHandoverAlgorithm`. Next, let's evaluate `A3RsrpHandoverAlgorithm` performance with the default `Hysteresis(3dB)` and `TimeToTrigger(256ms)` values in our LTE mobility scenario simulation by executing the following commands and observing the simulation results:

```
./ns3 run scratch/pkt-lte-handovers - "2"
..
UE location:221,0
UE location:173,0
UE location:211,0
/NodeList/0/DeviceList/0/LteEnbRrc/HandoverStartHandover Start
time:12.656 eNB CellId 1: start handover of UE with IMSI 19
RNTI 16 to CellId 2 Time:+1.2656e+10ns
/NodeList/1/DeviceList/0/LteEnbRrc/HandoverEndOkHandover End
time:12.6602 eNB CellId 2: completed handover of UE with IMSI
19 RNTI 29 Time:+1.26602e+10ns
UE location:178,0
UE location:201,0
..
..
```

```
UE location:213,0
UE location:131,0
/NodeList/1/DeviceList/0/LteEnbRrc/HandoverStartHandover Start
time:20.656 eNB CellId 2: start handover of UE with IMSI 22
RNTI 21 to CellId 1 Time:+2.0656e+10ns
/NodeList/0/DeviceList/0/LteEnbRrc/HandoverEndOkHandover End
time:20.6602 eNB CellId 1: completed handover of UE with IMSI
22 RNTI 29 Time:+2.06602e+10ns
UE location:218,0
UE location:121,0
```

From the handover start and end traces, let's track the following important details related to UE-18 and UE-21:

- UE-18 (IMSI 19) started handing over from eNB-1 (CellID-1) to eNB-2 (CellID-2) at 1.2656e+10ns and the handover completed at 1.2666e+10ns on eNB-2

- The handover delay for UE-18 is (1.2666e+10ns) –(1.2656e+10ns)

- UE-21 (IMSI 22) started handing over from eNB-2 (CellID-2) to eNB-1 (CellID-1) at 2.0656e+10ns and the handover completed at 2.06602e+10ns on eNB-1

- The handover delay for UE-21 is (2.06602e+10ns) – (2.0656e+10ns)

- We can also observe UEs' locations (X,Y) during the simulation execution

Now, let's change Hysteresis to 2 dB and TimeToTrigger to 100 ms when executing pkt-lte-handovers.cc. While executing the simulation, we can observe the following details:

```
./ns3 run scratch/pkt-lte-handovers - "2"
..
UE location:163,0
UE location:231,0
/NodeList/0/DeviceList/0/LteEnbRrc/HandoverStartHandover Start
time:10.7 eNB CellId 1: start handover of UE with IMSI 19 RNTI
16 to CellId 2 Time:+1.07e+10ns
/NodeList/1/DeviceList/0/LteEnbRrc/HandoverEndOkHandover End
time:10.7042 eNB CellId 2: completed handover of UE with IMSI
19 RNTI 29 Time:+1.07042e+10ns
UE location:168,0
UE location:221,0
..
..
```

```
UE location:208,0
UE location:141,0
/NodeList/1/DeviceList/0/LteEnbRrc/HandoverStartHandover Start
time:19.7 eNB CellId 2: start handover of UE with IMSI 22 RNTI
21 to CellId 1 Time:+1.97e+10ns
/NodeList/0/DeviceList/0/LteEnbRrc/HandoverEndOkHandover End
time:19.7042 eNB CellId 1: completed handover of UE with IMSI
22 RNTI 29 Time:+1.97042e+10ns
UE location:213,0
UE location:131,0
```

From these results, we can track the following important details related to UE-18 and UE-21 from the handover traces:

- UE-18 (IMSI 19) started handing over from eNB-1 (CellID-1) to eNB-2 (CellID-2) at 1.07e+10ns and the handover completed at 1.07042e+10ns on eNB-2. Here, we can note that the handover start is earlier compared to the previous evaluation. This is due to a decrease in the TimeToTrigger value.

- The handover delay for UE-18 is (1.07042e+10ns) –(1.07e+10ns). There is no change in the handover delay.

- UE-21 (IMSI 22) started handing over from eNB-2 (CellID-2) to eNB-1 (CellID-1) at 1.97e+10ns and the handover completed at 1.97042e+10ns on eNB-1. UE-21 handover from eNB-2 also started earlier due to a decrease in TimeToTrigger values.

- The handover delay for UE-21 is (1.97042e+10ns) – (1.97e+10ns). There is no change in the handover delay.

To conclude, the change in TimeToTrigger and Hysteresis led to a delay in the handover starting time. In this scenario, there is no change in throughput. But in other scenarios, throughput results might be affected. Hence, it is necessary to configure suitable handover thresholds to optimize handover performance. Excellent! We have completed the successful simulation of LTE mobility scenarios set up using ns-3 supporting handover algorithms.

## Summary

In this chapter, we practiced ns-3-supporting LTE important features such as radio resource scheduling for efficient spectrum management, ensuring QoS requirements using EPS bearers and the PSS scheduler, and mobility use-case scenario evaluations.

Mainly, we discussed how to evaluate various ns-3 supporting schedulers such as RR, PF, CQA, and PSS performance using interesting hands-on activities. Specifically, we discussed how to set up and test ns-3-supporting LTE QoS features using EPS bearers. Besides, as part of the LTE mobility scenarios' setup and evaluation, we discussed how to configure and evaluate `A2A4RsrqHandover` and `A3RsrpHandover` algorithms. As part of the evaluations, we discussed the importance of handover thresholds and their effect on mobility scenarios.

After exploring ns-3 LTE-supporting basic simulations, in the next chapter, users are going to learn how to set up LTE advanced networks such as HetNets, how to conduct site surveys for LTE networks' deployment, and how to manage interference in LTE networks using ns-3 LTE module supporting features.

# 9

# Researching LTE Advanced Networks: LTE HetNets and Interference Management Using ns-3

In this chapter, we will start by discussing an important LTE operator activity called site surveying, and planning for deploying eNBs or small cells (home eNBs) using heatmaps. Site surveying is a very important activity for mobile operators to maximize cell coverage and identify poor signal spots and black spots (no-signal areas). To conduct a site survey, the ns-3 LTE module offers **Radio Environment Maps** (**REMs**). In ns-3, to simulate LTE networks, users can generate REM plots and tune the operating power parameters and antennas of eNBs. REM plot insights about signal coverage help deploy eNBs and the optimal number of home eNBs (femtocells or small cells).

As part of the LTE advanced network evolution, home eNBs and small cells play a key role in improving cell coverage and bringing services closer to subscribers to provide better throughput results. Deploying eNBs and small cells together is known as a **heterogeneous network** (**HetNet**) **deployment**. In this chapter, we will discuss how to simulate HetNets in ns-3 using a hands-on activity. We will mainly discuss important ns-3 simulation aspects to consider when deploying home eNBs and how to configure home eNB operating modes to restrict their access and provide internet access.

Although mobile operators do site surveying and plan deployments for eNBs, UE transmissions still suffer from interference in the network. Besides careful site planning, operators have to optimally utilize the spectrum available. This forces operators to go for frequency-reuse one deployments (sharing the same spectrum with neighboring eNBs). Due to reuse one deployment scenarios, cell edge UE faces high interference from neighbor eNBs (or cells). Hence, it is necessary to manage spectrum resources to minimize interference on the network. As part of interference management, we will work through a hands-on activity to configure and test ns-3 supporting spectrum reuse algorithms to minimize interference on the network.

We will discuss the following topics in this chapter:

- Using ns-3 REM plots for site surveying and LTE deployments

- Setting up LTE HetNets using ns-3

- Exploring and configuring the LTE interference management algorithms supported in ns-3

# Using ns-3 REM plots for site surveying and LTE deployments

ns-3 REM plots are helpful for visualizing the signal coverage of eNBs (or cells) in a given deployment area. Using a REM plot, we can visualize the following important details:

- Usually, a REM plot is represented as three regions: an orange or red color region represents good signal strength, a blue region represents an average signal strength, and a blacked-out region represents very poor signal strength or no signal

- Based on the eNB's location, antenna orientation, and transmission power, a REM plot gives details on the respective eNB's coverage regions

- A REM plot has an *x*-axis and a *y*-axis to track eNB and UE deployment locations

Using REM plot colors and *x/y*-coordinates, we can identify signal coverage areas and black spots for eNBs. This helps operators to tune eNBs' antennas ad transmission power to improve their coverage. It also helps operators to decide in which regions small cells or home eNBs should be deployed. In this section, we will discuss the following hands-on activity to quickly understand how to conduct site surveys for deploying LTE eNBs and small cells. We mainly consider the following tasks as part of this hands-on activity:

- Deploying a three-sector eNB. That means we are going to deploy three cells.

- Randomly deploying UE around the eNB in an open space.

- Tracking the UE attachment details of the connected cells.

- Generating a REM plot and visualizing eNB coverage regions and UE locations.

- Inspecting cell coverage regions and identifying any black spots.

- Tuning the eNB's antennas to improve cell signal coverage.

- If there are any black spots, improving cell coverage using necessary small cell deployments.

- After deploying small cells, conducting site surveying again using a REM plot.

Without hesitation, in the next section, let's see how to implement this hands-on activity using ns-3 in a `pkt-lte-rem-survey.cc` simulation.

## LTE deployment site surveying using ns-3 – hands-on activity

Let's start by importing all of the following necessary packages:

```cpp
#include <ns3/core-module.h>
#include <ns3/network-module.h>
#include <ns3/mobility-module.h>
#include <ns3/internet-module.h>
#include <ns3/lte-module.h>
#include <ns3/config-store-module.h>
#include <ns3/buildings-module.h>
#include <ns3/point-to-point-helper.h>
#include <ns3/applications-module.h>
#include <ns3/log.h>
#include <iomanip>
#include <ios>
#include <string>
#include <vector>
using namespace ns3;
```

In order to generate a REM plot using the gnuplot tool, we need to track the eNB deployment locations in a file. Here, we use the following function to track the locations of the eNB and small cells in our simulation and save their respective locations in a given file:

```cpp
void
PrinteNbLocsToFile (std::string filename)
{
  std::ofstream outFile;
  outFile.open (filename.c_str (), std::ios_base::out |
std::ios_base::trunc);
  for (NodeList::Iterator it = NodeList::Begin (); it !=
NodeList::End (); ++it)
    {
      Ptr<Node> node = *it;
      int nDevs = node->GetNDevices ();
      for (int j = 0; j < nDevs; j++)
        {
          Ptr<LteEnbNetDevice> enbdev = node->GetDevice (j)-
>GetObject <LteEnbNetDevice> ();
```

```
        if (enbdev)
          {
            Vector pos = node->GetObject<MobilityModel>
()->GetPosition ();
            outFile << "set label \"" << enbdev->GetCellId ()
                    << "\" at "<< pos.x << "," << pos.y
                    << " left font \"Helvetica,4\" textcolor
rgb \"white\" front  point pt 2 ps 0.3 lc rgb \"white\" offset
0,0"
                    << std::endl;
          }
      }
    }
  }
}
```

Similarly, we also need to track the locations for all the UE in our simulation to plot these locations on the REM plot. We use the following function to save all the locations for all the pieces of UE in a file:

```
void
PrintUeLocsToFile (std::string filename)
{
  std::ofstream outFile;
  outFile.open (filename.c_str (), std::ios_base::out |
std::ios_base::trunc);
  for (NodeList::Iterator it = NodeList::Begin (); it !=
NodeList::End (); ++it)
    {
      Ptr<Node> node = *it;
      int nDevs = node->GetNDevices ();
      for (int j = 0; j < nDevs; j++)
        {
          Ptr<LteUeNetDevice> uedev = node->GetDevice (j)-
>GetObject <LteUeNetDevice> ();
          if (uedev)
            {
              Vector pos = node->GetObject<MobilityModel>
()->GetPosition ();
              outFile << "set label \"" << uedev->GetImsi ()
```

```
                           << "\" at "<< pos.x << "," << pos.y << "
left font \"Helvetica,4\" textcolor rgb \"grey\" front point pt
1 ps 0.3 lc rgb \"grey\" offset 0,0"
                           << std::endl;
              }
          }
      }
}
```

After implementing the functions to generate the REM plot, start main(), creating the necessary number of eNBs, home eNBs, and pieces of UE for our simulation activity:

```
int
main (int argc, char *argv[])
{
  uint16_t nUEs = 20;
  double simTime = 2.0;
  NodeContainer UEs;
  NodeContainer eNBs;
  NodeContainer heNBs;
  eNBs.Create(3);
  heNBs.Create(1);
  UEs.Create(nUEs);
```

Next, deploy a three-sector eNB (three cells) at a specific location. Then, randomly deploy UE around the eNB using ns3::RandomDiscPositionAllocator:

```
  Ptr<ListPositionAllocator> positionAlloc1 =
CreateObject<ListPositionAllocator> ();
  positionAlloc1->Add (Vector(300,300, 0));
  MobilityHelper mobility1;
  mobility1.
SetMobilityModel("ns3::ConstantPositionMobilityModel");
  mobility1.SetPositionAllocator(positionAlloc1);
  mobility1.Install(eNBs);

  mobility1.SetPositionAllocator
("ns3::RandomDiscPositionAllocator",
                           "X", StringValue ("300.0"),
```

```
                                    "Y", StringValue ("300.0"),
                                    "Rho", StringValue
("ns3::UniformRandomVariable[Min=0|Max=150]"));

    mobility1.Install(UEs);
```

Now, start configuring transmission power and antenna details for the eNB using `LteHelper`:

```
    NetDeviceContainer enbDevs;
    Ptr <LteHelper> lteHelper = CreateObject<LteHelper> ();
    Config::SetDefault ("ns3::LteEnbPhy::TxPower", DoubleValue
(46.0));
```

While configuring the three-sector eNB, we need to configure each antenna model, beamwidth, and gain separately. Let's configure the first cell's antenna orientation to 0 degrees, the beamwidth to 100 dB, and the gain to 0.0 using the following code snippet:

```
    lteHelper->SetEnbAntennaModelType
("ns3::CosineAntennaModel");
    //lteHelper->SetEnbAntennaModelAttribute ("Orientation",
DoubleValue (0));
    lteHelper->SetEnbAntennaModelAttribute ("Orientation",
DoubleValue (0));
    lteHelper->SetEnbAntennaModelAttribute
("HorizontalBeamwidth", DoubleValue (100));
    lteHelper->SetEnbAntennaModelAttribute ("MaxGain",
DoubleValue (0.0));
    enbDevs.Add ( lteHelper->InstallEnbDevice (eNBs.Get (0)));
```

Next, configure the second cell's antenna orientation to 120 degrees, beamwidth to 100 dB, and the gain to 0 using the following code snippet:

```
    lteHelper->SetEnbAntennaModelType ("ns3::CosineAntennaModel");
    //lteHelper->SetEnbAntennaModelAttribute ("Orientation",
DoubleValue (360/3));
    lteHelper->SetEnbAntennaModelAttribute ("Orientation",
DoubleValue (180));
    lteHelper->SetEnbAntennaModelAttribute
("HorizontalBeamwidth", DoubleValue (100));
    lteHelper->SetEnbAntennaModelAttribute ("MaxGain",
DoubleValue (0.0));
    enbDevs.Add ( lteHelper->InstallEnbDevice (eNBs.Get (1)));
```

Finally, configure the third cell's antenna orientation to 270 degrees, the beamwidth to 100 dB, and the gain to 0 using the following code snippet:

```
LteHelper->SetEnbAntennaModelType
("ns3::CosineAntennaModel");
  //lteHelper->SetEnbAntennaModelAttribute ("Orientation",
DoubleValue (2*360/3));
  lteHelper->SetEnbAntennaModelAttribute ("Orientation",
DoubleValue (270));
  lteHelper->SetEnbAntennaModelAttribute
("HorizontalBeamwidth", DoubleValue (100));
  lteHelper->SetEnbAntennaModelAttribute ("MaxGain",
DoubleValue (0.0));
  enbDevs.Add (lteHelper->InstallEnbDevice (eNBs.Get (2)));
```

After configuring and setting up the three-sector LTE eNB, install the LTE UE. Then, attach the UE to the closest eNB using the following code snippet:

```
NetDeviceContainer UEdevs = lteHelper->InstallUeDevice (UEs);
lteHelper->AttachToClosestEnb (UEdevs, enbDevs);
```

We just completed setting up a three-sector eNB and the UE for our simulation setup. Next, let's see how to set up an LTE home eNB (small cell). In ns-3, either an eNB or home eNB can be created using `NodeContainer`. As we initially created a node for setting up a home eNB, now we deploy it at a specific location, and configure its antenna mode and transmission power details using the following code snippet (initially, we comment the following home eNB deployment code):

```
/*  positionAlloc1->Add (Vector(600,500,0));
  mobility1.Install(heNBs);
  Config::SetDefault ("ns3::LteEnbPhy::TxPower", DoubleValue
(20.0));
  lteHelper->SetEnbAntennaModelType
("ns3::IsotropicAntennaModel");
  enbDevs.Add (lteHelper->InstallEnbDevice (heNBs.Get (0)));*/
```

We have just completed writing the necessary code for our simulation setup. Next, let's see how to generate a REM plot for our simulated deployment. In ns-3, a REM plot is generated using `RadioEnvironmentMapHelper`. To generate the REM plot, first, we need to track the location details for the eNB and UE in specific files. Here, we collect the eNB's locations in `enbs.txt` and the UE location details in `ues.txt`.

Then, we need to configure the `RadioEnvironmentMapHelper`-specific attributes, such as `OutputFile` for the REM and the boundaries of the simulation setup area in terms of XMin, XMax, YMin, YMax, and Z. We should also set the LTE DownLinkSpectrumChannel attribute. We have configured the aforementioned attributes using the following code snippet:

```
Ptr<RadioEnvironmentMapHelper> remHelper;
PrinteNbLocsToFile ("enbs.txt");
PrintUeLocsToFile ("ues.txt");
remHelper = CreateObject<RadioEnvironmentMapHelper> ();
remHelper->SetAttribute ("Channel", PointerValue (lteHelper-
>GetDownlinkSpectrumChannel ()));
remHelper->SetAttribute ("OutputFile", StringValue ("pkt-lte-
survey.rem"));
remHelper->SetAttribute ("XMin", DoubleValue (0));
remHelper->SetAttribute ("XMax", DoubleValue (700));
remHelper->SetAttribute ("YMin", DoubleValue (0));
remHelper->SetAttribute ("YMax", DoubleValue (700));
remHelper->SetAttribute ("Z", DoubleValue (1.5));
remHelper->Install ();
Simulator::Run ();
Simulator::Destroy ();
return 0;
}
```

Finally, we close `main()` by installing the REM helper and cleaning up the simulation setup resources at end of the simulation.

Well done. We have set up the simulation topology for the LTE deployment survey. In the next section, let's inspect the REM plots for our simulation setup topology.

## REM plot generation and inspection

Let's execute our simulation setup written in `pkt-lte-rem-survey.cc` by executing the following commands at our ns-3 installation directory:

```
./ns3 run scratch/pkt-lte-rem-survey
```

> **Important note**
>
> When we use the REM plot installation code in our simulation setup, your simulation stops automatically after generating the REM-related output files. No traffic simulation events will be executed. Hence, after generating REM plot output files, we have to comment the REM plot installation code and execute our simulation setup again to get the application's traffic simulation results.

It generates an output file called `pkt-lte-survey.rem`. This contains our simulation REM plot values. Next, let's see how to use it for generating a REM plot using the `gnuplot` tool.

We use the following `gnuplot` script (`remgrscript`) to generate our simulation setup REM plot:

```
set terminal postscript eps enhanced color font 'Helvetica,18'
linewidth 3.5
set output "survey.eps"
set view map;
set xlabel "X"
set ylabel "Y"
set cblabel "SINR (dB)"
unset key
plot "pkt-lte-survey.rem" using ($1):($2):(10*log10($4)) with
image
```

Next, execute the following command to generate the REM plot (shown in *Figure 9.1*) in the `survey.eps` output file:

**`gnuplot enbs.txt ues.txt remgrscript`**

Let's open `survey.eps` and we can observe the following details. *Figure 9.1* displays a REM plot for our simulation setup with three-sector eNB regions in three colors: orange, blue, and black. On the right-hand side of the REM plot, we can clearly observe each color representing the signal strength (the SINR in dB). Based on the current simulation setup and its three-antenna configuration and transmission power, we can spot three orange color spots that are related to the coverage areas of the three cells of the simulation setup's eNB (located at (300, 300) in *Figure 9.1*). In the REM plot, white dots represent our simulation setup UE locations (**X/Y**):

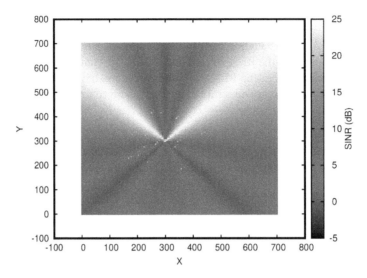

Figure 9.1 – A REM plot of our simulation setup

By zooming in on the plot (refer to *Figure 9.2*), we can clearly see the IMSIs of the UE and their regions. For example, UE **6** is experiencing a good SINR (>20 dB), UE **4** is experiencing an average SINR (>15 dB), and UE **1** is experiencing a poor SINR (<5 dB):

Figure 9.2 – Zooming in on our simulation setup REM plot

From the REM plot, we can also identify black spots. In *Figure 9.1*, black regions are nothing but black spots (areas with a very poor SINR).

In our current simulation setup, compared to cell-1 and cell-2, the cell-3 coverage area is smaller. Let's see the coverage of cell-3 when changing its antenna orientation from 270 to 240 degrees (change it in `pkt-lte-rem-survey.cc`). Then, execute the simulation setup and generate the REM plot again using the following commands in your simulation directory:

```
./ns3 run scratch/pkt-lte-rem-survey
gnuplot enbs.txt ues.txt remgrscript
```

Let's open the `survey.eps` file again to check the updated REM plot (shown in *Figure 9.3*):

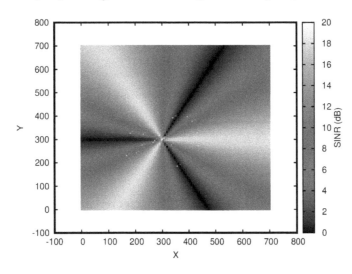

Figure 9.3 – The REM plot after changing cell-3's orientation from 270 to 240 degrees

From the updated REM plot (shown in *Figure 9.3*), we can clearly observe that now the coverage areas for all three cells are almost the same. This tells us that it is necessary to tune the antenna parameters to increase cell coverage.

From the updated REM plot (shown in *Figure 9.3*), we can also observe that there are black spots. Now, let's deploy a small cell (a home eNB) at a specific location and check the coverage regions for eNBs and home eNBs using a REM plot.

We can enable small cells in our simulation setup by uncommenting the home eNB deployment code. Then, we execute the following command:

```
./ns3 run scratch/pkt-lte-rem-survey
```

Next, generate a new REM plot *(Figure 9.4)* in `survey.eps` using the following command:

```
gnuplot enbs.txt ues.txt remgrscript
```

From *Figure 9.4*, it is clearly evident that at a home eNB location (400,400), the UE is experiencing a better SINR (we can see this by zooming in on the updated REM plot shown in *Figure 9.5*):

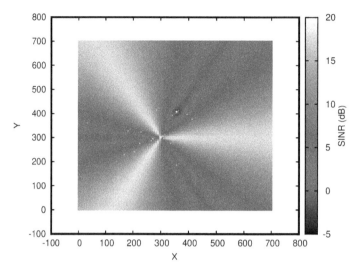

Figure 9.4 – REM plot showing the eNB and home eNB coverage regions

Let's zoom in on *Figure 9.4* and observe the home eNB coverage region in *Figure 9.5*:

Figure 9.5 – A REM plot showing home eNB coverage regions

Usually, a home eNB's operating transmission power (<20 dBm) is much lower than that of a macro eNB (46 dBm). Hence, its coverage area is also smaller, as shown in *Figure 9.5*. As we deployed a home eNB in one of the black-spot regions, we now see the UE facing good signal strengths. Similarly, operators can deploy home eNBs as necessary to improve coverage.

Well done. We have understood how to generate a REM plot for our simulation setup and inspect coverage regions, black spots, UE locations, and suitable locations for home eNB deployments. Next, we discuss an interesting topic – how to set up LTE advanced networks such as HetNets using ns-3.

## Setting up LTE HetNets using ns-3

As we discussed in the previous topic, even in open spaces, eNBs are not able to cover a complete area. Hence, it is necessary for operators to deploy small cells to improve the coverage area and serve their users with improved signal strengths. In this section, we will discuss how to set up LTE HetNets to meet indoor UE or home UE requirements (which are deployed inside building rooms), as well as outdoor UE requirements (outside building blocks). To set up LTE HetNets, we will complete the following tasks in our hands-on activity:

- Setting up building placement models
- Deploying home eNBs, such as femtocells, inside the building rooms
- Configuring home eNBs with **Closed Subscriber Group (CSG)** IDs to restrict their access to indoor UE
- Deploying macro eNBs (outdoor eNBs) to serve outdoor UE
- Providing all UE with internet access
- Testing the ns-3 CSG implementation details, such as restricting home eNBs' access
- Testing how home eNBs improve UE's data rates
- In the absence of home eNBs, testing how indoor UE is able to access LTE services

To carry out these tasks, we will implement our simulation setup in `pkt-lte-hetnets.cc`. Let's start implementing it in the following section.

## LTE HetNets setup in ns-3

Let's start implementing our simulation setup by importing the following necessary packages:

```
#include <ns3/core-module.h>
#include <ns3/network-module.h>
#include <ns3/mobility-module.h>
#include <ns3/internet-module.h>
#include <ns3/lte-module.h>
#include <ns3/config-store-module.h>
#include <ns3/buildings-module.h>
#include <ns3/point-to-point-helper.h>
#include <ns3/applications-module.h>
#include "ns3/flow-monitor-module.h"
#include <ns3/log.h>
#include <iomanip>
#include <ios>
#include <string>
#include <vector>
using namespace ns3;
```

To track the UE connection details for an eNB or home eNB, we will use the following callback function to print the UE connection details:

```
void
UEsConnectionStatus (std::string context,
                                 uint64_t imsi,
                                 uint16_t cellid,
                                 uint16_t rnti)
{
  std::cout << " UE IMSI " << imsi
          << " connected to CellId " << cellid
          << " with RNTI " << rnti<<"
Time:"<<Simulator::Now()
          << std::endl;
}
```

In order to create building models and deploy home eNBs (e.g., femtocells), we adopted the ns-3 FemtoBlockAllocator class to implement the following HomeNbBlockAllocator class. Basically, the HomeNbBlockAllcator implementation considers deploying home eNBs in buildings. When deploying a home eNB, this takes care of not overlapping a home eNB block with its neighboring home eNB blocks:

```
bool AreOverlapping (Box a, Box b)
{
  return !((a.xMin > b.xMax) || (b.xMin > a.xMax) || (a.yMin >
b.yMax) || (b.yMin > a.yMax));
}


class HomeNbBlockAllocator
{
public:
  HomeNbBlockAllocator (Box area, uint32_t nApartmentsX,
uint32_t nFloors);
  void Create (uint32_t n);
  void Create ();
private:
  bool OverlapsWithAnyPrevious (Box box);
  Box m_area; ///< Area
  uint32_t m_nApartmentsX; ///< X apartments
```

```
    uint32_t m_nFloors; ///< number of floors
    std::list<Box> m_previousBlocks; ///< previous bocks
    double m_xSize; ///< X size
    double m_ySize; ///< Y size
    Ptr<UniformRandomVariable> m_xMinVar; ///< X minimum variance
    Ptr<UniformRandomVariable> m_yMinVar; ///< Y minimum variance
};
```

To allocate home eNBs only within an operator's coverage region, we need to configure the coverage area boundaries for macro eNBs and set up building blocks for home eNBs using `GridBuildingAllocator`. Based on the number of apartments and floors, this sets the allocation boundaries for home eNBs using the following code:

```
HomeNbBlockAllocator::HomeNbBlockAllocator (Box area, uint32_t
nApartmentsX, uint32_t nFloors)
    : m_area (area),
      m_nApartmentsX (nApartmentsX),
      m_nFloors (nFloors),
      m_xSize (nApartmentsX*10 + 20),
      m_ySize (70)
{
  m_xMinVar = CreateObject<UniformRandomVariable> ();
  m_xMinVar->SetAttribute ("Min", DoubleValue (area.xMin));
  m_xMinVar->SetAttribute ("Max", DoubleValue (area.xMax - m_
xSize));
  m_yMinVar = CreateObject<UniformRandomVariable> ();
  m_yMinVar->SetAttribute ("Min", DoubleValue (area.yMin));
  m_yMinVar->SetAttribute ("Max", DoubleValue (area.yMax - m_
ySize));
}
```

The following function helps to create the necessary number of home eNB blocks:

```
void
HomeNbBlockAllocator::Create (uint32_t n)
{
  for (uint32_t i = 0; i < n; ++i)
    {
      Create ();
```

```
        }
    }
```

When setting up a home eNB deployment building block, it is necessary to check that it is not deployed beyond an operator's coverage area boundaries and does not overlap with neighboring home eNB building blocks within that coverage area. Using the following function, we create a home eNB deployment building block and set its boundaries carefully based on the number of rooms and floors using the GridBuildingAllocator class:

```
void
HomeNbBlockAllocator::Create ()
{
  Box box;
  uint32_t attempt = 0;
  do
    {
      NS_ASSERT_MSG (attempt < 100, "Too many failed attempts
to position apartment block. Too many blocks? Too small
area?");
      box.xMin = m_xMinVar->GetValue ();
      box.xMax = box.xMin + m_xSize;
      box.yMin = m_yMinVar->GetValue ();
      box.yMax = box.yMin + m_ySize;
      ++attempt;
    }
  while (OverlapsWithAnyPrevious (box));
  NS_LOG_LOGIC ("allocated non overlapping block " << box);
  m_previousBlocks.push_back (box);
  Ptr<GridBuildingAllocator>  gridBuildingAllocator;
  gridBuildingAllocator = CreateObject<GridBuildingAllocator>
();
  gridBuildingAllocator->SetAttribute ("GridWidth",
UintegerValue (1));
  gridBuildingAllocator->SetAttribute ("LengthX", DoubleValue
(10*m_nApartmentsX));
  gridBuildingAllocator->SetAttribute ("LengthY", DoubleValue
(10*2));
  gridBuildingAllocator->SetAttribute ("DeltaX", DoubleValue
(10));
```

```
  gridBuildingAllocator->SetAttribute ("DeltaY", DoubleValue
(10));
  gridBuildingAllocator->SetAttribute ("Height", DoubleValue
(3*m_nFloors));
  gridBuildingAllocator->SetBuildingAttribute ("NRoomsX",
UintegerValue (m_nApartmentsX));
  gridBuildingAllocator->SetBuildingAttribute ("NRoomsY",
UintegerValue (2));
  gridBuildingAllocator->SetBuildingAttribute ("NFloors",
UintegerValue (m_nFloors));
  gridBuildingAllocator->SetAttribute ("MinX", DoubleValue
(box.xMin + 10));
  gridBuildingAllocator->SetAttribute ("MinY", DoubleValue
(box.yMin + 10));
  gridBuildingAllocator->Create (2);
}
```

When setting up home eNB building blocks in the previous function, the following function is useful for checking whether any home eNB building blocks are overlapping with neighboring home eNB blocks:

```
bool
HomeNbBlockAllocator::OverlapsWithAnyPrevious (Box box)
{
  for (std::list<Box>::iterator it = m_previousBlocks.begin ();
it != m_previousBlocks.end (); ++it)
    {
      if (AreOverlapping (*it, box))
        {
          return true;
        }
    }
  return false;
}
```

Good. We have written the necessary home eNB building block placement-related code. Let's use it and set up our simulation. Let's start our simulation setup with the necessary number of nodes for setting up eNBs and UE:

```
int
main (int argc, char *argv[])
```

```
{
  uint16_t nUEs = 20;
  double simTime = 2.0;
  NodeContainer UEs;
  NodeContainer eNBs;
  eNBs.Create(1);
  UEs.Create(nUEs);
```

Then, we will deploy the macro eNB at a specific location and also deploy its UE (outdoor UE) nearby horizontally using the following position allocator code:

```
Ptr<ListPositionAllocator> positionAlloc1 =
CreateObject<ListPositionAllocator> ();
  positionAlloc1->Add (Vector(300,300, 0));
  for (uint16_t i = nUEs/2; i < nUEs; i++)
    {
      positionAlloc1->Add (Vector(300+i, 300, 0));
    }
  MobilityHelper mobility1;
  mobility1.
SetMobilityModel("ns3::ConstantPositionMobilityModel");
  mobility1.SetPositionAllocator(positionAlloc1);
  mobility1.Install(eNBs);
  mobility1.Install(UEs);
```

Then, we define the operator's coverage area box with suitable boundaries using macroUeBox. These boundaries are necessary for creating home eNB blocks and deploying home eNBs. In this example, we created a macroUeBox coverage area (0,600,0,600) box using the following lines of code:

```
Box macroUeBox;
double ueZ = 1.5;
macroUeBox = Box (0, 600, 0, 600, ueZ, ueZ);
```

Now, we start by defining our simulation setup's home eNB deployment building blocks using the following lines of code:

```
uint32_t nBlocks = 1;
uint32_t nApartmentsX = 1;
uint32_t nFloors = 1;
```

Let's create our simulation setup's home eNB blocks using the following lines of code:

```
HomeNbBlockAllocator blockAllocator (macroUeBox,
nApartmentsX, nFloors);
  blockAllocator.Create (nBlocks);
  uint32_t nHomeEnbs = round (2 * nApartmentsX * nBlocks *
nFloors);
  uint32_t nHomeUes = round (nHomeEnbs * 4);
```

We just created those building blocks and estimated the number of home eNBs based on the number of Apartments, Blocks, and Floors. We also defined the amount of home UE based on the home eNBs. Next, let's create the necessary number of ns-3 nodes for setting home eNBs and home UE using the following lines of code:

```
NodeContainer homeEnbs;
  homeEnbs.Create (nHomeEnbs);
  NodeContainer homeUes1, homeUes2;
  homeUes1.Create (nHomeUes/2);
  homeUes2.Create (nHomeUes/2);
  NodeContainer homeUes = NodeContainer (homeUes1,homeUes2);
  MobilityHelper mobility;
  mobility.SetMobilityModel
("ns3::ConstantPositionMobilityModel");
```

After creating the necessary macro eNB, UE, and home eNB and home UE nodes, let's install the LTE protocols and set up a complete LTE EPC network for the macro eNB and its UE using the following code:

```
Ptr <LteHelper> lteHelper = CreateObject<LteHelper> ();
  Ptr<PointToPointEpcHelper> epcHelper =
CreateObject<PointToPointEpcHelper> ();
  lteHelper->SetEpcHelper (epcHelper);
  Config::SetDefault ("ns3::LteEnbRrc::SrsPeriodicity",
UintegerValue (80));
  Config::SetDefault ("ns3::LteEnbPhy::TxPower", DoubleValue
(46.0));
  NetDeviceContainer eNBdevs = lteHelper->InstallEnbDevice
(eNBs);
  NetDeviceContainer UEdevs = lteHelper->InstallUeDevice (UEs);
```

Next, set up the `Internet` node and connect it to LTE EPC using the following lines of code:

```
NodeContainer internetHost;
InternetStackHelper internet;
internetHost.Create (1);
internet.Install (internetHost);
Ptr<Node> ihost = internetHost.Get (0);
Ptr<Node> pgw = epcHelper->GetPgwNode ();
```

Define the `Internet` link bandwidth and configure the EPC nodes for internet access using the following lines of code:

```
 PointToPointHelper p2p;
  p2p.SetDeviceAttribute ("DataRate", DataRateValue (DataRate
("10Gb/s")));
  p2p.SetDeviceAttribute ("Mtu", UintegerValue (1500));
  p2p.SetChannelAttribute ("Delay", TimeValue (Seconds
(0.010)));
  NetDeviceContainer inetDevs = p2p.Install (pgw,ihost);
  Ipv4AddressHelper ip;
  ip.SetBase ("1.0.0.0", "255.0.0.0");
  Ipv4InterfaceContainer ipinterfs = ip.Assign (inetDevs);
  Ipv4Address hostaddr = ipinterfs.GetAddress (1);
  Ipv4StaticRoutingHelper iproute;
  Ptr<Ipv4StaticRouting> ihostroute  = iproute.GetStaticRouting
(ihost->GetObject<Ipv4> ());
  ihostroute->AddNetworkRouteTo (Ipv4Address ("7.0.0.0"),
Ipv4Mask ("255.0.0.0"), 1);
```

Similarly, configure the LTE UE for accessing the `Internet` node using the following lines of code:

```
internet.Install (UEs);
  Ipv4InterfaceContainer ueIPs;
  ueIPs = epcHelper->AssignUeIpv4Address (NetDeviceContainer
(UEdevs));
  for (uint32_t i = 0; i < UEs.GetN (); ++i)
    {
      Ptr<Node> ue = UEs.Get (i);
      Ptr<Ipv4StaticRouting> ueroute = iproute.GetStaticRouting
(ue->GetObject<Ipv4> ());
```

```
    ueroute->SetDefaultRoute (epcHelper-
>GetUeDefaultGatewayAddress (), 1);
    }
```

Attach the outdoor UE to the macro eNB using the following lines of code:

```
for (uint16_t i = 0; i < nUEs; i++)
    {
        lteHelper->Attach (UEdevs.Get(i), eNBdevs.Get(0));
    }
```

After completing this macro eNB and UE setup, we will set up the home eNBs and UE `Pathloss` models using the following line of code:

```
lteHelper->SetAttribute ("PathlossModel", StringValue
("ns3::HybridBuildingsPropagationLossModel"));
lteHelper->SetPathlossModelAttribute ("ShadowSigmaExtWalls",
DoubleValue (0));
lteHelper->SetPathlossModelAttribute ("ShadowSigmaOutdoor",
DoubleValue (1));
lteHelper->SetPathlossModelAttribute ("ShadowSigmaIndoor",
DoubleValue (1.5));
// use always LOS model
lteHelper->SetPathlossModelAttribute ("Los2NlosThr",
DoubleValue (1e6));
lteHelper->SetSpectrumChannelType
("ns3::MultiModelSpectrumChannel");
```

Next, we deploy the home eNBs only inside building regions using the following lines of code:

```
Ptr<PositionAllocator> positionAlloc =
CreateObject<RandomRoomPositionAllocator> ();
mobility.SetPositionAllocator (positionAlloc);
mobility.Install (homeEnbs);
BuildingsHelper::Install (homeEnbs);
```

After deploying the home eNBs, we deploy the home UE only inside building rooms using the following lines of code:

```
positionAlloc = CreateObject<SameRoomPositionAllocator>
(homeEnbs);
mobility.SetPositionAllocator (positionAlloc);
```

```
mobility.Install (homeUes);
BuildingsHelper::Install (homeUes);
```

After completing the home eNB and UE setup, we configure the home eNBs' transmission power, antenna model, and CSG ID (= 1) for restricting UE access. Then, we install the LTE protocol stack on the home eNBs using the following lines of code:

```
Config::SetDefault ("ns3::LteEnbPhy::TxPower", DoubleValue
(23.0));
  lteHelper->SetEnbAntennaModelType
("ns3::IsotropicAntennaModel");
  lteHelper->SetEnbDeviceAttribute ("CsgId", UintegerValue
(1));
  lteHelper->SetEnbDeviceAttribute ("CsgIndication",
BooleanValue (true));
  NetDeviceContainer homeEnbDevs = lteHelper->InstallEnbDevice
(homeEnbs);
```

Next, we configure the CSG ID (= 1) and installed the LTE protocol stack on home UE to access the home eNBs using the following lines of code. If we set the CSG ID to something other than 1, instead of the home UE attaching to the home eNBs, they will be attached to available macro eNBs:

```
lteHelper->SetUeDeviceAttribute ("CsgId", UintegerValue (1));
  NetDeviceContainer homeUeDevs1 = lteHelper->InstallUeDevice
(homeUes1);
  lteHelper->SetUeDeviceAttribute ("CsgId", UintegerValue (1));
  NetDeviceContainer homeUeDevs2 = lteHelper->InstallUeDevice
(homeUes2);
  NetDeviceContainer homeUeDevs = NetDeviceContainer
(homeUeDevs1, homeUeDevs2);
```

Similar to outdoor UE, the home UE is also configured with the Internet stack, and their routing tables for accessing the Internet node are set up using the following lines of code:

```
internet.Install (homeUes);
  Ipv4InterfaceContainer hueIPs;
  hueIPs = epcHelper->AssignUeIpv4Address (NetDeviceContainer
(homeUeDevs));
  for (uint32_t i = 0; i < homeUes.GetN (); ++i)
    {
      Ptr<Node> ue = homeUes.Get (i);
```

```
    Ptr<Ipv4StaticRouting> ueroute = iproute.GetStaticRouting
(ue->GetObject<Ipv4> ());
    ueroute->SetDefaultRoute (epcHelper-
>GetUeDefaultGatewayAddress (), 1);
    }
```

Attach the home UE to the closest home eNBs or macro eNB using the following lines of code:

```
lteHelper->Attach (homeUeDevs1);
lteHelper->Attach (homeUeDevs2);
```

After attaching the home UE, let's install the UDP DL flow applications to test the home UE's throughput:

```
uint16_t dlPort = 5001;
ApplicationContainer clientApps;
ApplicationContainer serverApps;
for (uint32_t u = 0; u < homeUes.GetN (); ++u)
  {
    PacketSinkHelper dlPacketSinkHelper
("ns3::UdpSocketFactory", InetSocketAddress
(Ipv4Address::GetAny (), dlPort));
    serverApps.Add (dlPacketSinkHelper.Install (homeUes.
Get(u)));
    UdpClientHelper dlClient (hueIPs.GetAddress (u), dlPort);
    dlClient.SetAttribute ("Interval", TimeValue
(Seconds(0.001)));
    dlClient.SetAttribute ("PacketSize", UintegerValue(512));
    dlClient.SetAttribute ("MaxPackets",
UintegerValue(10000));
    clientApps.Add (dlClient.Install (ihost));
    dlPort++;
  }
serverApps.Start (Seconds (1.01));
clientApps.Start (Seconds (1.01));
```

Similarly, let's install the UDP DL flow applications to test the macro eNB UE's throughput:

```
uint16_t dlPort1 = 6001;
ApplicationContainer eclientApps;
ApplicationContainer eserverApps;
```

```
    for (uint32_t u = 0; u < UEs.GetN (); ++u)
      {
        PacketSinkHelper dlPacketSinkHelper
  ("ns3::UdpSocketFactory", InetSocketAddress
  (Ipv4Address::GetAny (), dlPort1));
        serverApps.Add (dlPacketSinkHelper.Install (UEs.Get(u)));
        UdpClientHelper dlClient (ueIPs.GetAddress (u), dlPort);
        dlClient.SetAttribute ("Interval", TimeValue
  (Seconds(0.01)));
        dlClient.SetAttribute ("PacketSize", UintegerValue(512));
        dlClient.SetAttribute ("MaxPackets",
  UintegerValue(10000));
        clientApps.Add (dlClient.Install (ihost));
        dlPort1++;
      }
    eserverApps.Start (Seconds (1.01));
    eclientApps.Start (Seconds (1.01));
```

Configure and connect the UE's connection traces in our simulation to track the UE attachment details using the following lines of code:

```
  Config::Connect ("/NodeList/*/DeviceList/*/LteUeRrc/
  ConnectionEstablished",
                    MakeCallback (&UEsConnectionStatus));
```

Finally, let's close main () by installing a flow monitor to check the UE's throughput results and free the simulation resources:

```
  Ptr<FlowMonitor> flowmon;
  FlowMonitorHelper flowmonHelper;
  flowmon = flowmonHelper.InstallAll ();
  Simulator::Stop(Seconds(2.0));
  Simulator::Run ();
  flowmon->SerializeToXmlFile ("hetnet_flowmetrics.xml", true,
  true);
  Simulator::Destroy ();
  return 0;
}
```

Excellent! We have completed our LTE HetNet simulation setup using macro eNBs and home eNBs. Let's evaluate it thoroughly.

# LTE HetNet evaluation

Start evaluating the simulation setup by allowing all UE (indoor UE or outdoor UE) to connect to the macro eNB. To do this, we need to ensure that the home UE's CSG ID does not match the home eNBs' CSG ID. In our simulation setup, the home eNBs are configured with a CSG ID of 1. Hence, we need to set the home UEs' CSG ID to something other than 1 in pkt-lte-hetnet.cc. Then, we will execute our pkt-lte-hetnet.cc simulation using the following command in our ns-3 installation directory:

```
./ns3 run scratch/pkt-lte-hetnet | tee noheNBaccess
cat noheNBaccess |grep 'CellId 1'
...
UE IMSI 21 connected to CellId 1 with RNTI 49
Time:+2.60214e+08ns
UE IMSI 25 connected to CellId 1 with RNTI 46
Time:+2.60214e+08ns
UE IMSI 26 connected to CellId 1 with RNTI 48
Time:+2.60214e+08ns
UE IMSI 28 connected to CellId 1 with RNTI 47
Time:+2.60214e+08ns
UE IMSI 23 connected to CellId 1 with RNTI 55
Time:+2.65214e+08ns
UE IMSI 27 connected to CellId 1 with RNTI 54
Time:+2.65214e+08ns
UE IMSI 24 connected to CellId 1 with RNTI 53
Time:+2.65214e+08ns
UE IMSI 22 connected to CellId 1 with RNTI 100
Time:+7.63214e+08ns
cat noheNBaccess |grep 'CellId 1' |wc -l
28
```

Here, we collected the UE connection detail results in no_heNBs_access. By counting the amount of UE connected to CellId 1, we can confirm that all UE has connected to the macro eNB successfully.

Now, check all the UE throughput results by executing the following commands:

```
python3 scratch/flowmon-parse-results.py hetnet_flowmetrics.xml
> no_heNBs_access
cat no_heNBs_access |grep 'RX ' |tail -n 8
RX bitrate: 687.07 kbit/s
RX bitrate: None
```

```
RX bitrate: 684.93 kbit/s
RX bitrate: 686.36 kbit/s
RX bitrate: 675.70 kbit/s
RX bitrate: 683.29 kbit/s
RX bitrate: 550.09 kbit/s
RX bitrate: 634.50 kbit/s
```

We collected the indoor UE throughput results in the no_heNBs_access file. From the file, we displayed the last eight UE (which are home or indoor UE) throughput results. We can observe from the results that few indoor UE got zero throughput or lower throughput values compared to the throughput results for outdoor UE.

Next, let's enable access to indoor UE for the home eNBs by setting the same CSG ID for both the home eNBs and indoor UE. We configure a CSG ID to 1 for the home eNBs and indoor UE in our simulation setup (changing it in pkt-lte-hetnet.cc). Then, we execute our simulation setup with the following commands:

```
/ns3 run scratch/pkt-lte-hetnet | tee heNBaccess
cat heNBaccess |grep 'CellId 1' |wc -l
20
cat heNBaccess |grep 'CellId 2' |wc -l
4
cat heNBaccess |grep 'CellId 3' |wc -l
4
```

We collected the home eNB access results in the heNBaccess file. From the file, we can count the amount of UE connected to each of the home eNBs and the macro eNB. From the results, we can observe that out of eight pieces of home UE, four pieces of home UE are connected to home eNB-1, and another four pieces of home UE are connected to home eNB-2. All outdoor UE is connected to the macro eNB. Next, let's check all the UE throughput results. In particular, we will check the home UE throughput results by executing the following commands:

```
python3 scratch/flowmon-parse-results.py hetnet_flowmetrics.xml
> heNBs_access
cat heNBs_access |grep 'RX ' |tail -n 8
RX bitrate: 867.63 kbit/s
RX bitrate: 2645.72 kbit/s
RX bitrate: 876.72 kbit/s
RX bitrate: 885.80 kbit/s
RX bitrate: 401.44 kbit/s
```

```
RX bitrate: 409.26 kbit/s
RX bitrate: 1001.09 kbit/s
RX bitrate: 594.45 kbit/s
```

From the results, we can clearly observe that now the indoor UE throughput results have improved compared to connecting them to macro eNBs. Now, no indoor UE suffers from zero throughput. Moreover, some indoor UE experiences very good throughput results. From this hands-on activity, we have understood that HetNets are highly helpful to improve operators' cell coverage and capacity. In particular, HetNets help indoor UE to enjoy good signal strength and higher throughput results.

Next, we will discuss another important activity called interference management in LTE networks to efficiently manage the operating spectrum and ensure a fair throughput for edge UE.

## Exploring and configuring the LTE interference management algorithms supported in ns-3

Let's imagine that an operator purchases a 3 MHz spectrum from the government and decides to use it efficiently. Having a limited spectrum, operators must reuse the same spectrum for all their coverage regions. To do so efficiently, operators usually go for a frequency-reuse 1 deployment. This means an eNB (cell) and its neighboring eNBs (cells) use the same operating spectrum bandwidth (RBs). However, a frequency-reuse 1 deployment (all cells use the same frequency with fixed power – refer to *Figure 9.6*) leads to high interference from neighboring cells for edge UE (UE at cell boundaries) and edge UE suffers from very poor throughput:

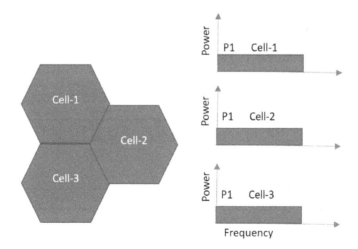

Figure 9.6 – Example of a frequency-reuse 1 deployment with a  fixed power plan

Hence, it is important to manage interference on the network and offer fair throughput results for UE in a cell. In research, a variety of fractional frequency reuse approaches with dynamic power control have evolved to manage interference in LTE networks as well as possible. As shown in *Figure 9.7*, the frequency is divided into sub-bandwidths (shown in the figure with different colors) and shared between cells with mutual exclusion. Moreover, each sub-bandwidth is accessed by cells using dynamic power control:

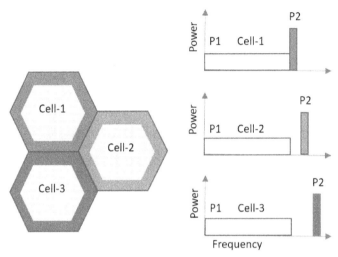

Figure 9.7 – Example of a fractional frequency reuse deployment with dynamic power control

In this section, we will discuss how to configure ns-3-supporting interference management algorithms and will test their performance using a hands-on activity. Before starting our hands-on activity, let's quickly come to understand the ns-3 interference management approaches.

To make our interference management discussion simple, let's visualize each cell operating spectrum of 3 MHz bandwidth (15 RBs) as follows:

Here's cell-1:

| 1 | 2 | 3 | 4 | 5 | 6 | 7 | 8 | 9 | 10 | 11 | 12 | 13 | 14 | 15 |
|---|---|---|---|---|---|---|---|---|----|----|----|----|----|----|

And cell-2:

| 1 | 2 | 3 | 4 | 5 | 6 | 7 | 8 | 9 | 10 | 11 | 12 | 13 | 14 | 15 |
|---|---|---|---|---|---|---|---|---|----|----|----|----|----|----|

And cell-3:

| 1 | 2 | 3 | 4 | 5 | 6 | 7 | 8 | 9 | 10 | 11 | 12 | 13 | 14 | 15 |
|---|---|---|---|---|---|---|---|---|----|----|----|----|----|----|

If the operator deploys 3 neighboring cells in a frequency-reuse 1 deployment, every cell can access all 15 RBs. This helps reuse the operating bandwidth efficiently. However, it results in severe interferences with edge UE (UE that overlaps with eNBs). If there is no interference management needed for eNBs, schedulers can choose any RBs to serve edge UE and make high-power transmissions. Then, neighboring eNBs may also use the same RBs to serve their edge UE at a high transmission power. Parallel transmissions from these eNBs using the same RBs will lead to high interference with the corresponding UE's reception. Hence, it is necessary to share edge UE RB allocation details from the scheduler at an eNB with neighbor eNBs. This helps neighbor eNBs schedulers allocate the same set of RBs to their UE with mutual exclusion to minimize interference. Because of the adoption of OFDMA technology in LTE standards, there is great flexibility in allocating individual RBs to UE based on the UE's channel conditions and interference details. This motivated LTE researchers to implement novel interference management algorithms for efficiently utilizing the operating spectrum and improving cell edge UE bit rates. Let's understand the following ns-3-supporting interference management algorithms quickly:

- By default, ns-3 LTE simulations run with frequency-reuse 1. This means all RBs are available to all eNBs in a simulation setup. In this setup, edge UE (in overlapping regions of cells) may suffer from interference from neighbor eNBs (or cells).

- **Strict Frequency Reuse** (**Strict FR**): This divides the available RBs into a reuse 1 band (for inner UE) and a reuse 3 band (for edge UE). In a cell, inner UE cannot access the reuse 3 band and edge UE cannot access the reuse 1 band. Strict FR classifies UE into inner UE or edge UE based on their RSRQ reports (similar to the handover measurement reports discussed in *Chapter 8*). For example, in a 3 MHz bandwidth, 15 RBs are divided between the reuse 1 band (1 to 9 RBs) and the reuse 3 band (10 to 15 RBs). Then, the reuse 1 band (1 to 9 RBs) is available to the inner UE of those cells only. Additionally, the reuse 3 band again is divided into 3 exclusive sub-bands (e.g., [10,11 RBs] to cell-1, [12,13 RBs] to cell-2, [14,15 RBs] to cell-2) for serving edge UE only. Strict FR follows a high transmission power plan for the reuse 3 band and a low transmission power plan for the reuse 1 band to minimize interference. Although it minimizes interference for edge UE, a cell can use a maximum of 11 RBs. Hence, this leads to lower spectrum utilization.

- **Soft Frequency Reuse** (**Soft FR**): Available RBs are also divided into a reuse 1 band and a reuse 3 band and follow low and high transmission power plans. To improve spectrum utilization (and use more RBs at a cell), for example, cell-1 can assign edge RBs of neighbor cells (cell-2 [12, 13 RBs] and cell-3 [14, 15 RBs]) to its inner UE at a low transmission power. On the other hand, Soft FR can also be implemented by giving a cell's inner UE access to edge RBs [10,11 RBs] when there is no edge UE. Definitely it can improve spectrum utilization, but it may again lead to interference.

- **Soft Fractional Frequency Reuse** (**Soft FFR**): This also divides available RBs into a reuse-1 band and a reuse-3 band. In order to improve spectrum utilization further, it offers three power plans by dividing cell UE into inner (low-power), middle (medium-power), and edge UE (high-power). It classifies UE based on its RSRQ reports with two RSRQ thresholds. With this approach, middle UE is served by the reuse 1 band only with a medium-power plan, edge UE is served with edge RBs only, and the inner UE of cell-1 can be served by cell-2 and cell-3 reuse-3 band RBs with a low-power plan. However, cell edge RBs cannot be used for inner UE. Hence, it improves the throughput for inner UE compared to the Strict FR and Soft FR approaches.

- **Enhanced FFR**: This also divides the available RBs into a reuse 1 band and reuse 3 band. It only follows low- and high-transmission power plans. In order to improve spectrum utilization in a cell when reuse 3 band RBs are not utilized by any edge UE, those RBs can be assigned to inner UE. Hence, it is possible to access all RBs from all cells, which improves spectrum utilization and controls interference.

Next, as part of a hands-on activity, we will discuss how to configure a few of these algorithms and test their working.

## ns-3 supporting LTE interference management approaches

We discuss a hands-on activity to learn how to use ns-3-supporting LTE interference management approaches. In order to implement this activity, we use the hands-on activity code for scheduler performance evaluation (`pkt-lte-scheduler.cc`) discussed in *Chapter 8*. We make the following necessary changes in `pkt-lte-scheduler.cc` to implement the following activities in `pkt-lte-interference-mgmt.cc`:

- Print log details of the interference management approaches to track edge UE details
- Set up suitable placement models for simulating high interference between two eNBs
- Configure Strict FR, Soft FR, and Enhanced FFR interference management approaches in our simulation
- Similar to `pkt-lte-scheduler.cc`, ensure all UE starts its downlink traffic from the internet simultaneously
- Evaluate how Strict FR, Soft FR, and Enhanced FFR work

Let's start implementing our hands-on activity in (`pkt-lte-interference-mgmt.cc`) by making the following modification to `pkt-lte-scheduler.cc`.

Include the following lines of code to enable log details for Strict FR, Soft FR, and Enhanced FFR approaches at the start of `main()`. This helps you trace various UE RSRQ details, edge UE details, and RB allocation details:

```
LogComponentEnable ("LteFfrEnhancedAlgorithm", LOG_LEVEL_
INFO);
LogComponentEnable ("LteFrStrictAlgorithm", LOG_LEVEL_ALL);
LogComponentEnable ("LteFfrSoftAlgorithm", LOG_LEVEL_ALL);
```

Next, to simulate an interference scenario, keep half of the UE for each cell in an overlapping region between the cells (two eNBs) using the following lines of code instead of the `pkt-lte-scheduler.cc` placement code:

```
Ptr<ListPositionAllocator> positionAlloc =
CreateObject<ListPositionAllocator> ();
  positionAlloc->Add (Vector(0, 0, 0));
  positionAlloc->Add (Vector(500, 0, 0));
  for (uint16_t i = 0; i < nUEs/2; i++)
    {
      if (i>10)
      {
        positionAlloc->Add (Vector(i+300, 0, 0));
      }
      else
        positionAlloc->Add (Vector(i+10, 0, 0));
    }
  for (uint16_t i = nUEs/2; i < nUEs; i++)
    {
      if (i<30)
        positionAlloc->Add (Vector(i+200, 0, 0));
      else
        positionAlloc->Add (Vector(i+500, 0, 0));
    }
```

After deciding the UE locations, install the placement and mobility models for all UE and eNBs:

```
MobilityHelper mobility;
mobility.
SetMobilityModel("ns3::ConstantPositionMobilityModel");
  mobility.SetPositionAllocator(positionAlloc);
  mobility.Install(eNBs);
  mobility.Install(UEs);
```

Then, after the scheduler configuration code, include the following lines of code to configure the Strict FR, Soft FR, and Enhanced FFR algorithms in our simulation setup. We pass the algorithm name as a second command-line argument during execution. Let's first configure the Strict FR algorithm using the following lines of code:

```
lteHelper->SetFfrAlgorithmType(argv[2]);
std::string frAlgorithmType = lteHelper->GetFfrAlgorithmType
```

```
();
  if (frAlgorithmType == "ns3::LteFrStrictAlgorithm")
    {
       lteHelper->SetFfrAlgorithmAttribute ("RsrqThreshold",
UintegerValue (25));
    }
```

Here, we used all default parameters of ns3::LteFrStrictAlgorithm and configured the RSRQ threshold as 25 for detecting edge UE. That means UE experiencing less than an RSRQ value of 25 is considered edge UE.

Next, configure the Soft FFR algorithm using the following lines of code. Here, we configure two RSRQ thresholds for detecting center UE and edge UE and transmission power plans for the center, middle, and edge areas using the following lines of code:

```
    else if (frAlgorithmType == "ns3::LteFfrSoftAlgorithm")
    {
      lteHelper->SetFfrAlgorithmAttribute
("CenterRsrqThreshold", UintegerValue (30));
      lteHelper->SetFfrAlgorithmAttribute ("EdgeRsrqThreshold",
UintegerValue (25));
      lteHelper->SetFfrAlgorithmAttribute
("CenterAreaPowerOffset",
                                            UintegerValue
(LteRrcSap::PdschConfigDedicated::dB_6));
      lteHelper->SetFfrAlgorithmAttribute
("MediumAreaPowerOffset",
                                            UintegerValue
(LteRrcSap::PdschConfigDedicated::dB_1dot77));
      lteHelper->SetFfrAlgorithmAttribute
("EdgeAreaPowerOffset",
                                            UintegerValue
(LteRrcSap::PdschConfigDedicated::dB3));
      lteHelper->SetFfrAlgorithmAttribute ("CenterAreaTpc",
UintegerValue (1));
      lteHelper->SetFfrAlgorithmAttribute ("MediumAreaTpc",
UintegerValue (2));
      lteHelper->SetFfrAlgorithmAttribute ("EdgeAreaTpc",
UintegerValue (3));
    }
```

Finally, configure the Enhanced FFR algorithm using the following lines of code with its default parameter and an RSRQ threshold as 25 for detecting edge UE:

```
else if (frAlgorithmType == "ns3::LteFfrEnhancedAlgorithm")
  {
    lteHelper->SetFfrAlgorithmAttribute("RsrqThreshold",
UintegerValue (25));
  }
```

Then, replace the `lteHelper->InstallEnbDevice(eNBs)` line with the following lines of code in `pkt-lte-interference-mgmt.cc`:

```
NetDeviceContainer eNBdevs;
lteHelper->SetFfrAlgorithmAttribute ("FrCellTypeId",
UintegerValue (1));
eNBdevs.Add (lteHelper->InstallEnbDevice (eNBs.Get (0)));
lteHelper->SetFfrAlgorithmAttribute ("FrCellTypeId",
UintegerValue (2));
eNBdevs.Add (lteHelper->InstallEnbDevice (eNBs.Get (1)));
```

> **Important note**
> Similar to `pkt-lte-scheduler.cc` in *Chapter 8*, here, also we used the same UDP traffic configuration lines of code to simulate simultaneous UE downlink traffic from the internet.

That's all. We use the same UDP client and server lines of code from `pkt-lte-scheduler.cc` to run our `pkt-lte-interference-mgmt.cc` simulation.

Next, let's evaluate our simulation setup and test how the Strict FR, Soft FFR, and Enhanced FFR algorithms work.

## Testing how the ns-3-supporting LTE Strict FR, Soft FFR, and Enhanced FFR algorithms work

Let's start by executing our interference simulation setup with the Strict FR algorithm using the following commands:

```
script strict
./ns3 run scratch/pkt-lte-interference-mgmt -- "2"
"ns3::LteFrStrictAlgorithm"
exit
```

At end of the simulation, it generates `strict` (for tracking the edge UE details) and `lte_sched.xml` files. Let's parse `lte_sched.xml` to get the UE throughput results (into `strict.csv`) using the following commands:

```
python3 scratch/flowmon-parse-results.py lte_sched.xml >
strictres
cat strictres |grep 'RX ' |cut -f3 -d ' ' |tail -40 | tail -40
> strict.scv
```

Next, execute our interference simulation setup with the Soft FFR algorithm using the following commands:

```
script ffrsoft
./ns3 run scratch/pkt-lte-interference-mgmt -- "2"
"ns3::LteFfrSoftAlgorithm"
exit
```

At end of the simulation, it generates `ffrsoft` (for tracking edge UE details) and `lte_sched.xml` files. Let's parse `lte_sched.xml` to get the UE throughput results (into `soft.csv`) using the following commands:

```
python3 scratch/flowmon-parse-results.py lte_sched.xml >
ffrsoftres
cat ffrsoftres |grep 'RX ' |cut -f3 -d ' ' |tail -40 | tail -40
> soft.scv
```

Finally, execute our interference simulation setup with the Enhanced FFR algorithm using the following commands:

```
Script effr
./ns3 run scratch/pkt-lte-interference-mgmt -- "2"
"ns3::LteFfrEnhancedAlgorithm"
exit
```

At end of the simulation, it generates `effr` (for tracking the edge UE details) and `lte_sched.xml` files. Let's parse `lte_sched.xml` to get the UE throughput results (into `effr.csv`) using the following commands:

```
python3 scratch/flowmon-parse-results.py lte_sched.xml >
ffrsoftres
cat ffrsoftres |grep 'RX ' |cut -f3 -d ' ' |tail -40 | tail -40
> soft.scv
```

Let's combine the Strict FR, Soft FFR, and Enhance FFR algorithms throughput results into a single file for plotting using the following command:

```
paste strict.csv soft.csv effr.csv >compareffralgo.csv
```

Then, generate a comparison plot of the Strict FR, Soft FFR, and Enhanced FFR interference management results using `compareffralgo.csv`. We generated the following plot using **LibreOffice**:

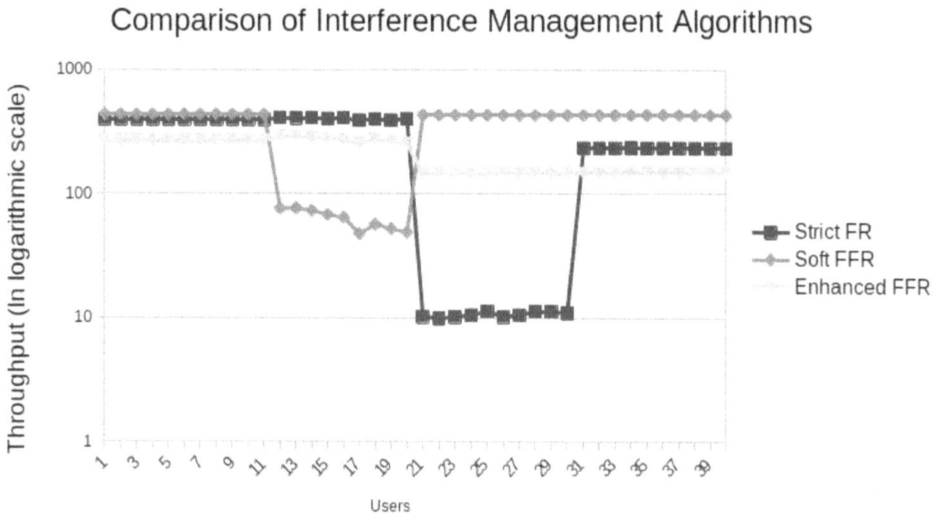

Figure 9.8 – Comparison of ns-3-supporting interference algorithms

Before inspecting the plot results, let's inspect the interference management algorithm logs to understand the results. Let's execute the following commands to see which UE is served using edge band RBs:

```
cat strict |grep 'Edge ' |cut -f3 -d ' ' |sort >strictRNTI
cat ffrsoft |grep 'Edge ' |cut -f3 -d ' ' |sort >softRNTI
cat effr |grep 'Edge ' |cut -f3 -d ' ' |sort >effrRNTI
```

Here, we collected the edge UE RNTI details into algorithm-specific output files. Next, we find the complete details for the edge UE, such as serving cell IDs and IMSIs, using the following commands:

```
cat strict | grep 'IMSI'  | cut -f3,4,7,8,10,11 -d '
'>StrictUEIMSI
cat ffrsoft | grep 'IMSI'  | cut -f3,4,7,8,10,11 -d '
'>SoftUEIMSI
cat effr | grep 'IMSI'  | cut -f3,4,7,8,10,11 -d ' '>EffrUEIMSI
```

Finally, to display the complete details for the edge UE of a given interference management algorithm, we use the following Python script (edgeUEdetails.py):

```python
import sys
rntilist = []
a = open(sys.argv[1],'r')
rntilist = a.readlines()
UEdetails = []
b = open(sys.argv[2],'r')
UEdetails = b.readlines()
for rnti in rntilist:
        for UE in UEdetails:
                if rnti in UE:
                        print(UE)
```

For example, let's inspect the details for the Strict FR serving edge UE:

```
Python3   edgeUEdetails.py strictRNTI StrictUEIMSI
IMSI 14 CellId 1 RNTI 103
IMSI 15 CellId 1 RNTI 111
IMSI 13 CellId 1 RNTI 128
IMSI 17 CellId 1 RNTI 133
IMSI 19 CellId 1 RNTI 15
IMSI 34 CellId 2 RNTI 15
IMSI 16 CellId 1 RNTI 56
IMSI 25 CellId 2 RNTI 156
IMSI 18 CellId 1 RNTI 67
IMSI 20 CellId 1 RNTI 78
IMSI 12 CellId 1 RNTI 97
```

Similarly, we can observe the Soft FFR and Enhance FFR algorithm logs. From the Strict FR log, we observe that most of the edge UE belongs to cell-1 only. By default, for an ns-3 Strict FR implementation, there is no reuse 3 band allocated to any edge UE. As we run with default configuration, most of cell-1's edge UE (11 to 20) and only a few pieces of cell-2's edge UE (21 to 30) are served using an edge RB band (observed in the log). Hence, cell-2's edge UE (21 to 30) suffers from poor throughput results as shown in *Figure 9.8*. We suggest users change DlEdgeSubBandwidth and inspect the results as part of their learning.

If comparing Soft FFR with Strict FFR, cell-1's edge UE (11 to 20) experiences a better throughput. However, comparing cell-1's edge UE (21 to 30) with cell-2's edge UE, cell-2's edge UE experiences a better throughput. This is due to Soft FFR trying to utilize all RBs for all UE (including inner and edge UE), resulting in some edge UE suffering from lower throughput results due to interference (shown in *Figure 9.8*). However, this is a promising improvement in the spectrum utilization compared to Strict FR and Enhanced FFR.

Finally, with Enhanced FFR, we observe fair throughput results for all UE (shown in *Figure 9.8*). However, the overall spectrum utilization is lower compared to Strict FFR and Soft FFR. The main reason for this is that Enhanced FFR allows inner UE to access reuse 3 band RBs only if no edge UE accesses those RBs. Hence, it promises fairness for all edge UE at the cost of lower spectrum utilization. Overall, we can clearly understand that when designing interference management algorithms, it is necessary to look into the edge UE throughput results as well as maximizing spectrum utilization.

> **Important note**
>
> This is a simple scenario for understanding LTE interference management and using ns-3-supporting interference management algorithms. We suggest users change the edge UE locations in the setup and the default parameters of the Strict FR, Soft FFR, and Enhanced FFR algorithms to inspect more interesting results.

Excellent! We have learned how to set up an LTE interference simulation setup and evaluate ns-3-supporting interference management algorithms.

# Summary

In this chapter, we learned how to simulate advanced LTE features, such as conducting site surveys, setting up HetNets, and evaluating interference management algorithms using the ns-3 LTE module. Although this chapter did not cover advanced LTE topics in depth, it gave a quick understanding of important details, such as REM plots, the importance of small cells in HetNets and how to configure their access, and the importance of interference management algorithms to offer fair throughput results to edge UE. In particular, we discussed in detail how to generate a REM plot and understand its details for site survey, how to simulate a realistic HetNet setup using building models, and how to manage the spectrum to control interference in an LTE network using strict frequency reuse, fractional frequency reuse, and enhanced fractional frequency reuse techniques.

Well done! We would like to thank you for showing interest in this book and learning about advanced network simulations using ns-3. We hope you enjoyed reading this book and learning about basic network simulations, Wi-Fi technologies, and LTE networks using ns-3. We recommend users carefully go through all the hands-on activities in this book and execute the simulations to get the most out of it. We wish you all the best!

# Index

www.packtpub.com

Subscribe to our online digital library for full access to over 7,000 books and videos, as well as industry leading tools to help you plan your personal development and advance your career. For more information, please visit our website.

## Why subscribe?

- Spend less time learning and more time coding with practical eBooks and Videos from over 4,000 industry professionals

- Improve your learning with Skill Plans built especially for you

- Get a free eBook or video every month

- Fully searchable for easy access to vital information

- Copy and paste, print, and bookmark content

Did you know that Packt offers eBook versions of every book published, with PDF and ePub files available? You can upgrade to the eBook version at packtpub.com and as a print book customer, you are entitled to a discount on the eBook copy. Get in touch with us at customercare@packtpub.com for more details.

At www.packtpub.com, you can also read a collection of free technical articles, sign up for a range of free newsletters, and receive exclusive discounts and offers on Packt books and eBooks.

# Other Books You May Enjoy

If you enjoyed this book, you may be interested in these other books by Packt:

**Mastering Python for Networking and Security - Second Edition**

José Manuel Ortega

ISBN: 9781839217166

- Create scripts in Python to automate security and pentesting tasks
- Explore Python programming tools that are used in network security processes
- Automate tasks such as analyzing and extracting information from servers
- Understand how to detect server vulnerabilities and analyze security modules
- Discover ways to connect to and get information from the Tor network
- Focus on how to extract information with Python forensics tools

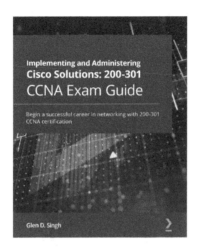

**Implementing and Administering Cisco Solutions: 200-301 CCNA Exam Guide**

Glen D. Singh

ISBN: 9781800208094

- Understand the benefits of creating an optimal network
- Create and implement IP schemes in an enterprise network
- Design and implement virtual local area networks (VLANs)
- Administer dynamic routing protocols, network security, and automation
- Get to grips with various IP services that are essential to every network
- Discover how to troubleshoot networking devices

## Packt is searching for authors like you

If you're interested in becoming an author for Packt, please visit authors.packtpub.com and apply today. We have worked with thousands of developers and tech professionals, just like you, to help them share their insight with the global tech community. You can make a general application, apply for a specific hot topic that we are recruiting an author for, or submit your own idea.

## Share your thoughts

Now you've finished *Advanced Network Simulations Simplified*, we'd love to hear your thoughts! Scan the QR code below to go straight to the Amazon review page for this book and share your feedback or leave a review on the site that you purchased it from.

https://packt.link/r/1804614459

Your review is important to us and the tech community and will help us make sure we're delivering excellent quality content.

# Download a free PDF copy of this book

Thanks for purchasing this book!

Do you like to read on the go but are unable to carry your print books everywhere?

Is your eBook purchase not compatible with the device of your choice?

Don't worry, now with every Packt book you get a DRM-free PDF version of that book at no cost.

Read anywhere, any place, on any device. Search, copy, and paste code from your favorite technical books directly into your application.

The perks don't stop there, you can get exclusive access to discounts, newsletters, and great free content in your inbox daily

Follow these simple steps to get the benefits:

1. Scan the QR code or visit the link below

https://packt.link/free-ebook/978-1-80461-445-7

2. Submit your proof of purchase

3. That's it! We'll send your free PDF and other benefits to your email directly

www.ingramcontent.com/pod-product-compliance
Lightning Source LLC
Chambersburg PA
CBHW062053050326
40690CB00016B/3072